I0062889

COOL CAREERS WITHOUT COLLEGE FOR
ANIMAL
LOVERS

COOL CAREERS WITHOUT COLLEGE

COOL CAREERS WITHOUT COLLEGE FOR
ANIMAL
LOVERS

**CHRIS
HAYHURST**

The Rosen Publishing Group, Inc.
New York

Published in 2002 by The Rosen Publishing Group, Inc.
29 East 21st Street, New York, NY 10010

First Edition

Library of Congress Cataloging-in-Publication Data

Hayhurst, Chris.
Animal lovers / Chris Hayhurst.
p. cm. — (Cool careers without college)
Includes bibliographical references (p.).
Summary: Profiles the characteristics of and qualifications needed for twelve jobs that involve working with animals.
ISBN: 978-1-4358-8811-1
1. Animal specialists—Vocational guidance—United States—juvenile literature. [1. Animal specialists—Vocational guidance. 2. Vocational guidance.] I. Title. II. Series.
SF80 .H39 2001
636'.023'73—dc21

2001003170

Manufactured in the United States of America

CONTENTS

INTRODUCTION

Do you love animals? If the answer is yes, then here's some good news: The working world—where most of us end up before too long—is full of jobs for animal lovers. Jobs in veterinary clinics, animal shelters, and pet stores, in the country or in the city, indoors or outdoors, are there for the taking, and, best of all, many of these jobs don't require a college degree. In fact, some require nothing more than a high-school diploma, the desire to help and care for animals, and a good work ethic. If you're a hard worker, there's a job out there for you.

The truth is, if animals are your thing, your future looks bright. Each year more and more jobs in animal care, animal protection, animal control, and other related fields become available to those who want them. Take a look at the jobs profiled in this book. Some—like pet shop worker or veterinary technician—you may have heard of. Others—like equine sports massage therapist or assistance-dog trainer—may be new, at least to you. Think about what it takes to get each job, and what each job is like. If you decide you're interested in learning more about a specific career, make use of the directory section at the end of each chapter. There you'll find other books and magazines on the subject, handy Web sites, and a list of professional associations offering more information. Take the time to explore your options. One day, perhaps not long from now, you'll land the job of your dreams—working with animals.

1

ASSISTANCE-DOG TRAINER

If you're lucky enough to have a pet dog in your family, you probably already have some experience in this field. Assistance dogs are work dogs. They act as assistants— as helpers or guides—for disabled people. But they're not born with the ability to obey their owners and behave properly. They must be taught dozens, sometimes hundreds of commands.

Teaching dogs to help people is what professional assistance-dog trainers do.

Description

People with many different kinds of mental and physical disabilities use assistance dogs. Blind people use Seeing Eye dogs, which are trained to guide them along sidewalks, streets, and hallways while avoiding dangerous obstacles like other people, cars, walls, and open doors. Deaf people might use assistance dogs to recognize warning sounds like fire alarms. People who are extremely shy or afraid to go out in public might keep an assistance dog by their side to gain the confidence they need to face the world.

Teaching dogs to be useful—to do more than normal dog activities like sleep, run, and eat—is very difficult. But with a bag full of special techniques and skills, lots of practice, and plenty of patience, you'll find that a career as an assistance-dog trainer is also very rewarding. You'll not only grow close to your canine friends, but you'll also meet and work with amazing people. The dogs are only

Tim Cordes, who is legally blind, walks onto the stage during his graduation ceremony with his Seeing Eye dog. He is the second completely blind person to be admitted to medical school.

Physically challenged people usually form very strong bonds of attachment to their assistance dogs.

half of your job; the other half is teaching your clients—those with the disabilities—how to use your trainees.

You have to love dogs to be an assistance-dog trainer. In many cases you'll be around them—anywhere from one to a dozen—every day, every night, and every weekend. Round-the-clock training is the most effective way to teach a dog how to work, so there's a good chance the dogs may live with you at your home. You'll spend most of your days teaching them all the useful tricks their future owners will need them to know—things like how to pick up keys off the floor,

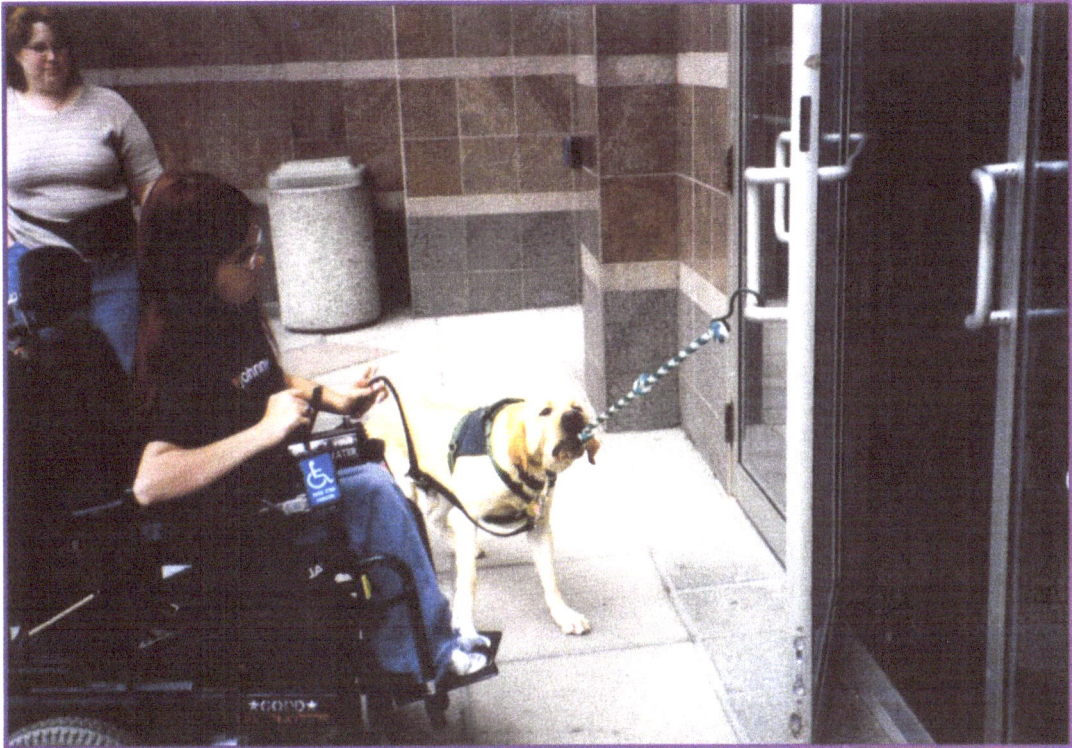

Assistance dogs can help the wheelchair bound perform everyday tasks.

how to turn on the lights using a wall switch, and how to close a door. You'll use a teaching technique called positive reinforcement. Positive reinforcement is a way of rewarding dogs for doing what they're told instead of punishing them for what they do wrong. You may teach them to pull a wheelchair. You will also teach them to sit quietly at their owners' feet during meals, to refrain from barking or fighting with other dogs, and to stop at intersections to let traffic pass before leading their owners across busy roads. Of course, you'll also get to play with them. They are dogs, after all!

Here, an assistance dog has learned how to pick up and carry a package for his young owner.

Education and Training

For now, anyway, formal education or certification is not required to become an assistance-dog trainer. The main qualifications are that you have to be good with dogs and you have to be good with people.

The best way to learn how to train assistance dogs is through hands-on experience. If there's a nonprofit dog-training business in your area—even if it's just a regular pet-training facility—see if you can volunteer your time or can work as an apprentice. Offer to help out on weekends or on weekdays after school. By spending time with professional trainers and learning from them, you'll be a step ahead of the game when it comes to starting your career.

Another good way to get into the field and to see if it's something you really want to do is by volunteering to work with disabled people. You'll learn to understand them and their needs and to have a better idea why assistance dogs are so important in their lives.

Outlook

Each year more and more people realize just how important a role dogs can play in the lives of disabled people. You've probably even seen "working dogs" yourself; they're easy to spot, with their bright vests and professional looks. Assistance dogs are now allowed on airplanes, in restaurants, in stores; anywhere their owners go, they can go, too. With the growing acceptance of assistance dogs, and the increased demand for them, the job market for professional trainers looks great. Still, don't enter this field for the money—the truth is, salaries are very low. Because most assistance-dog training companies are nonprofit organizations, they have very little money to go around and must struggle to raise the funds required to buy the dogs and raise and train them. Employees are there for the love of their work and the desire to help people, not to get rich.

Karen, an assistance-dog trainer for Assistance Dogs of the West in Santa Fe, New Mexico, has this to say about her work:

The dogs are always learning. That's why they live with me. If I leave the bathroom door open, they push the door shut for me. If I drop the keys at the front door, they pick them up for me. If I'm feeding, they have to wait. If I'm eating dinner, they need to be under the table and not begging. If I'm cooking, they can't be sitting right under the stove.

The dogs are a wonderful part of the business. It's incredible to work with them and teach them, but I think what it's really about is the clients—the people with the disabilities. Teaching the disabled to work with the dogs is more important than training the dogs. And to watch the dogs go work with somebody who really needs them is touching to me. I don't know how else I could be rewarded for my work.

Karen's Typical Day

2:00 AM

Get up and let one or two dogs outside so they can go to the bathroom. Go back to sleep.

5:30 AM

Get up, put bark collars on the dogs that bark, wake up the dogs that don't want to get up, and send everyone outside.

7:00 AM

Breakfast. Make each dog sit and wait for its meal to be served. When they finish, it's back outside.

8:30 AM

Time to work. Bring one or two dogs at a time back inside to work on new tricks and practice old ones.

10:30 AM

All the dogs go into the truck, and it's off to the office. Do office work for an hour while the dogs wait in the yard outside.

11:30 AM

Back into the truck for a trip to the local high school. Students take the dogs and learn what it takes to train them and work with them. The dogs learn to be comfortable around strangers and to obey commands from people other than their regular trainer.

3:15 PM

Return to the office for more paperwork.

4:00 PM

Load up the dogs for the last time, and return home.

4:30 PM

Pull out one or two dogs from the bunch and work with them individually.

5:30 PM

Dinner!

FOR MORE INFORMATION

BOOKS

Burch, Mary. *How Dogs Learn.* Saint Paul, MN: Hungry Mind Press, 1999.

Marlo, Shelby. *Shelby Marlo's New Art of Dog Training: Balancing Love and Discipline.* Lincolnwood, IL: NTC/Contemporary Publishing Co., 1999.

Owens, Paul. *The Dog Whisperer: A Compassionate, Nonviolent Approach to Dog Training.* Holbrook, MA: Adams Media Corp., 1999.

Pryor, Karen. *Don't Shoot the Dog! How to Improve Yourself and Others through Behavioral Training.* New York: Simon & Schuster, 1984.

Spector, Morgan. *Clicker Training for Obedience: Shaping Top Performance Positively.* Waltham, MA: Sunshine Books, 1999.

Tillman, Peggy. *Clicking with Your Dog, Step-by-Step in Pictures.* Waltham, MA: Sunshine Books, 2000.

WEB SITES
The American Dog Trainers Network
http://www.canine.org
A great resource for dog owners and trainers.

Clickertraining.com
http://www.clickertraining.com

Working Dogs Cyberzine
http://www.workingdogs.com

MAGAZINES
APDT Newsletter
Association of Pet Dog Trainers
17000 Commerce Parkway, Suite C
Mt. Laurel, NJ 08054
Bimonthly newsletter with latest training techniques and other helpful information.

Bulletin on Companion Animal Behavior
ABCS Inc.
2288 Manning Avenue
Los Angeles, CA 90064

Interactions
Delta Society
289 Perimeter Road East
Renton, WA 98055-1329
Quarterly magazine for members of the Delta Society.

VIDEOS

Clicker Magic!
Produced by Karen Pryor.
This video includes twenty demonstrations by top dog trainers working with young and old dogs. One of the best videos on the market. Running time: 55 minutes.

ASSOCIATIONS

Assistance Dogs International Inc.
c/o Canine Partners for Life
334 Faggs Manor Road
Cochranville, PA 19330
(610) 869-4902
Web site: http://www.assistance-dogs-intl.org.

Association of Pet Dog Trainers
17000 Commerce Parkway, Suite C
Mt. Laurel, NJ 08054
(800) 738-3647
Web site: http://www.apdt.com.

Canine Companions for Independence
P.O. Box 446
Santa Rosa, CA 95402-0446
(425) 226-7357
(800) 572-2275
Web site: http://www.caninecompanions.org.

Delta Society
289 Perimeter Road East
Renton, WA 98055-1329
(800) 869-6898
Web site: http://www.deltasociety.org.

International Association of Assistance Dog Partners
38691 Filly Drive
Sterling Heights, MI 48310
(810) 826-3938
Web site: http://www.iaadp.org

National Association of Dog Obedience Instructors
Attn: Corresponding Secretary, PMB #369
729 Grapevine Highway, Suite 369
Hurst, TX 76054-2085
Web site: http://www.nadoi.org.

2

FARMER

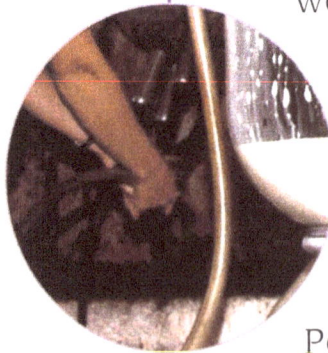

When most people imagine life as a farmer, they picture a lonely man working long hours in endless fields, picking corn, hoeing potatoes, and driving tractors. What they don't realize is that many farmers also work with animals. Dairy farmers work with cows. Poultry farmers work with chickens. Other farmers spend their days tending sheep, pigs, goats, turkeys, bees, and many other animals. It's true—grains and vegetables are a big part of the farming lifestyle, but so are animals. So come on! Pull on your boots, cinch up your belt, and beat a path for the barn!

Description

Farming is definitely no walk in the park. Most farmers work hard all of their lives and still barely eke out a living. Still, farming has its perks—especially if you like animals. Farmers who work with animals can specialize in many different disciplines. For example, livestock and cattle farmers raise cows for milk and meat. They feed the cattle with corn and hay, carc for them when they get sick, and provide shelter for them when the weather is bad. If the cows are dairy cows, they must be milked twice a day—once early in the morning and once at night. The milk must be cleaned and purified and then shipped to a bottling company so it can be sold to consumers. After the cows are milked, the farmer spends time "mucking" the milking stalls, washing away the cow paddies and sweeping out dust and dirt. The milking equipment must also be cleaned and prepared for the next run.

Poultry farmers work with chickens. They raise chicks until they grow into adults. The hens are then ready to lay

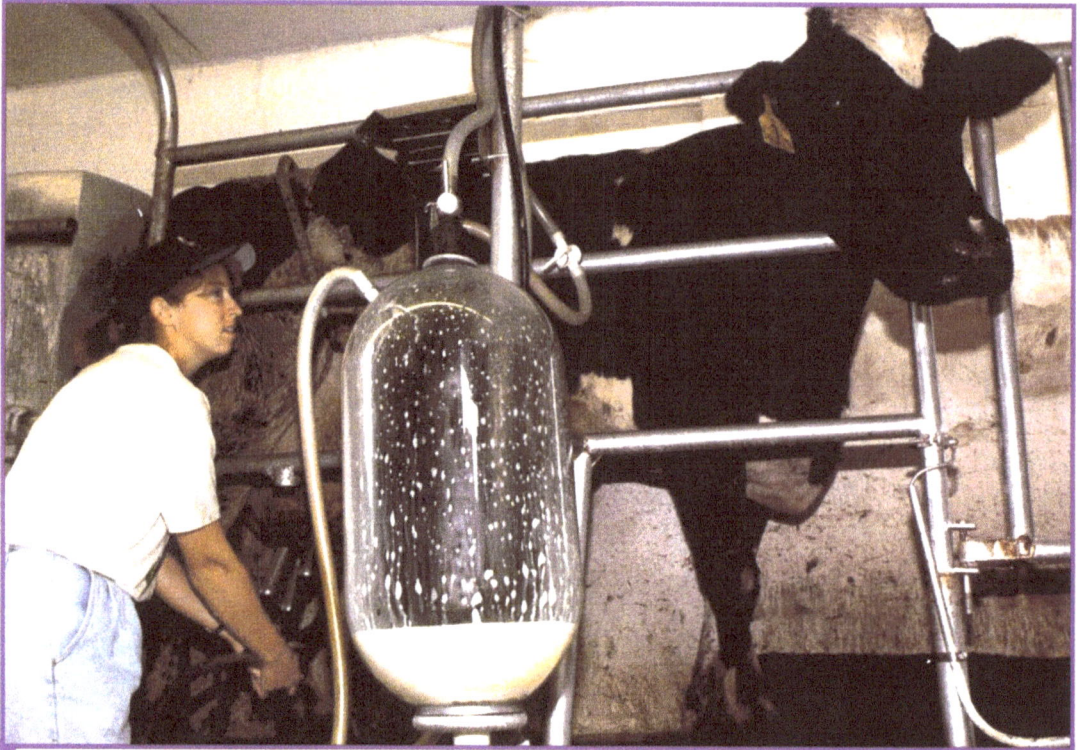

A Tennessee farmer adjusts the milking machine at her dairy farm. The metal device over the cow's hips prevents the animal from kicking.

eggs and the roosters can be sold for meat. On "free range" farms, farmers let them run and move about in spacious fenced-off areas. When the hens lay their eggs, the farmer collects them and sells them. Other animal farmers work with sheep. They raise the sheep and shear their wool for use in making sweaters, blankets, and other items. Still others work as beekeepers. They sell honey collected from the beehives and lend bees to other farmers who use them to pollinate crops.

A beekeeper shows his granddaughter how to collect honey from a honeycomb. They wear protective clothing to prevent them from being stung.

Whatever animals the farmer works with, the job often involves the same basic tasks. Animals must be fed and given water. Stalls and pens must be cleaned and fences and equipment must be maintained. Animals must be loaded onto trucks, unloaded into pastures, and herded into barns or cages. They must be watched closely for injuries and diseases, and helped when it's time to give birth. If an animal becomes sick, the farmer is the first to know and must do what he can to help it. He might treat it himself, or, if it's serious, he might call a veterinarian to come in and do the work.

These days, the farmer's workday rarely ends when the sun goes down. Most farmers now use computers to help manage their businesses and spend time every day (or night) updating their records and checking on sales and expenses. Farm machinery—like tractors and hay balers—must be serviced and repaired. Buildings require upkeep. If a farmer has just a few employees or no employees at all, it's easy to imagine how hard all of this work must be. Still, most farmers wouldn't trade their jobs for anything in the world. For them, making an honest living off the land while working with animals is worth the effort.

Education and Training

There was a time when almost all farmers grew up on farms. They were naturals, ready to work the land as soon as they were big enough to hold a shovel. Today, however, you don't have to be raised in a farming family to become a farmer. All you need is a willingness to work hard and learn on the job. Other ways to gain experience before you even get out of school are by volunteering or apprenticing. Offer your services to a farm in your area. They probably won't be able to pay

Sheep shearing is a physically demanding job, and the shearer must be careful to avoid causing serious cuts.

you much, if anything, but it's a safe bet they could use the extra help and can teach you all about the job.

Another good source of experience is through youth educational programs like 4-H and the National Future Farmers of America Organization. These groups teach kids the basics of farming and working with livestock. You can learn how to show animals at fairs and how to raise pigs and cows for sale; you'll also learn about the business side of farming. And speaking of business, be sure to take lots of math and computer classes in addition to any animal science courses you can get your hands on in high school. You'll need to be a solid number cruncher and know how to keep accurate records when it comes time to run your own business!

Outlook

According to the U.S. Bureau of Labor Statistics, in 1998 there were 851,000 professional farmers in the United States, and many of these farmers worked in dairy and livestock production. Unfortunately, the agency believes that number will get smaller in the next ten years, mainly because farming is such a difficult occupation at which to succeed. Many farmers lose money every year. Those that do make a profit can expect to make between $20,000 and $45,000 per year.

There is good news, however. As long as people must eat to survive, the world will need farmers. So as old farmers retire, young farmers—like you—can step in and take their places. Also, there's an interesting movement in the works: organic farming. With organic farming, the farmer does not use chemicals and lets the animals live as free and healthy a life as possible. Many experts believe that farmers who provide organic products—including organically produced milk from cows—stand an excellent chance of succeeding. People like to buy products that do not harm the environment. That could be good for your business, and even better news for the earth!

In Action

Most high-school students spend their summers doing typical summer jobs—things like waiting tables, lifeguarding, mowing lawns, or baby-sitting. Not Doug Smillers. A high-school sophomore in a small town in Nebraska, Doug earns money doing everything from helping his neighbors move cattle to selling llamas. His main job, however—the one he finds himself at most often—is at a local dairy. The dairy has 600 head of cattle, most of which need milking twice a day. "Sometimes there's no rhyme or reason for when they need me," says Doug. "They just call and I go to work." His eight- to ten-hour shifts usually take place early in the morning or late at night. He milks, feeds, calves—just about anything and everything a full-time, professional dairy farmer would do. In fact, he works so hard he's developed a bit of a reputation. "People say I have a good work ethic," says Doug. "And that makes me feel great."

FOR MORE INFORMATION

BOOKS

Benson, Laura Lee. *Organic Dairy Farming.* Gay Mills, WI: Orang-utan Press, 1996.

Ekarius, Carol. *Small-Scale Livestock Farming: A Grass-Based Approach for Health, Sustainability, and Profit.* Pownal, VT: Storey Books, 1999.

MAGAZINES

Bee Culture
Subscription Dept., Dept. W
623 West Liberty Street
Medina, OH 44256
(800) 289-7668 ext. 3255
Web site: http://bee.airoot.com/beeculture/index.htm

The Community Farm: A Voice for Community Supported Agriculture
3480 Potter Road
Bear Lake, MI 49614
(616) 889-3216

Sheep! **Magazine**
P.O. Box 10
Lake Mills, WI 53551
(414) 648-8285
Web site: http://www.sheepmagazine.com

Small Farm Today
Missouri Farm Publishing, Inc.
3903 West Ridge Trail Road
Clark, MO 65243-9525
(573) 687-3148
(800) 633-2535
Web site: http://www.smallfarmtoday.com

The Western Dairyman
Dept. WM
P.O. Box 819
Corona, CA 92878-0819
(909) 735-2730
Web site: http://www.dairybusiness.com

ASSOCIATIONS

Alternative Farming Systems Information Center
National Agricultural Library USDA
10301 Baltimore Avenue, Room 304
Beltsville, MD 20705-2351
(301) 504-6559
Web site: http://www.nal.usda.gov/afsic.

American Farm Bureau Federation
225 Touhy Avenue
Park Ridge, IL 60068
(847) 685-8600
Web site: http://www.fb.org.

American Society of Farm Managers and Rural Appraisers
950 South Cherry Street, Suite 508
Denver, CO 80246-2664
(303) 758-3513
Web site: http://www.asfmra.org

Center for Rural Affairs
101 South Tallman Street
P.O. Box 406
Walthill, NE 68067
(402) 846-5428
Web site: http://www.cfra.org.

National 4-H Headquarters
CSREES/USDA, Stop 2225
1400 Independence Avenue SW
Washington, DC 20250-2225
(202) 720-2908
Web site: http://www.4H-USA.org.

National Future Farmers of America
P.O. Box 68960
6060 FFA Drive
Indianapolis, IN 46268-0960
(317) 802-6060
Web site: http://www.ffa.org.

U.S. Department of Agriculture
Higher Education Program
1400 Independence Avenue SW, STOP 2251
Washington, DC 20250
(202) 720-1973
Web site: http://www.usda.gov

EQUINE SPORTS MASSAGE THERAPIST

If you've got a way with horses, and you're also good with your hands, there's a job for you. Equine sports massage therapists massage horses. They use their hands and arms to rub horses' muscles. By rubbing the muscles in special ways in specific places, they can help horses recover from

33

An equine massage therapist limbers the hind leg of a seven-month-old filly.

injuries, warm down and relax after a race or competition, and perform better in the future.

Description

As an equine sports massage therapist, you'll most certainly be working outside year-round. You'll spend a lot of time in barns and at horse shows. You'll be around horses all the time. If you like horses, this may be one of the best jobs in the world. The work is tough. It's physically demanding, often exhausting, after a full day of work. Horses are big animals, with big muscles. Rubbing these muscles requires good technique, but it also requires some strength. But don't worry—if you're not strong now, you will be with practice.

The only equipment you'll need is a sturdy stool to help you reach high spots on big horses, and your hands and arms. You'll spend an hour or more on each horse, working on problem areas and responding to their subtle reactions to your touch. The owner might be there with you, explaining how the horse feels when she rides it and how she hopes you can help through your massage.

Equine sports massage therapists are almost always self-employed. That means when they're through giving massages for the day, their job is not over. They have to spend time running their businesses—taking care of billing, making phone calls, doing paperwork. They distribute their business cards to local tackle shops and feed stores and go to jumping competitions and other equine events to do demonstrations and to pick up new clients. The equine sports massage therapist spends just as much time finding work as she does doing work, at least in the beginning.

As business picks up, the equine sports massage therapist might find that more and more of her work comes from referrals. Horse owners that know she does a good job might recommend her to other horse owners, and so on down the line. Eventually she might get so busy that she can pick and choose which horses she will and will not work on; she can specialize, say, in rodeo horses or jumpers. At this point, she'll know she's succeeded. She's making a living working with horses!

Education and Training

The first step on the road to becoming a professional equine sports massage therapist is learning to become comfortable with horses. Spend as much time around as many different horses as you can, and learn to ride them, groom them, and care for them. Become familiar with how they move and how they react to your movements. Learn what it takes to get close to a horse without spooking it, or making it nervous.

You can learn all of these skills by working at a local stable. At first you might have to volunteer your time, cleaning stalls or brushing down horses that have just been ridden. Ask any trainers and riders you work around lots of questions. The more you inquire, the more you'll learn. Later, if you're lucky, you might be paid for your help. Either way, the experience you gain will help you later. As you gain horse experience, you should also read books about

A woman sponges the nostrils of a horse as part of its grooming.

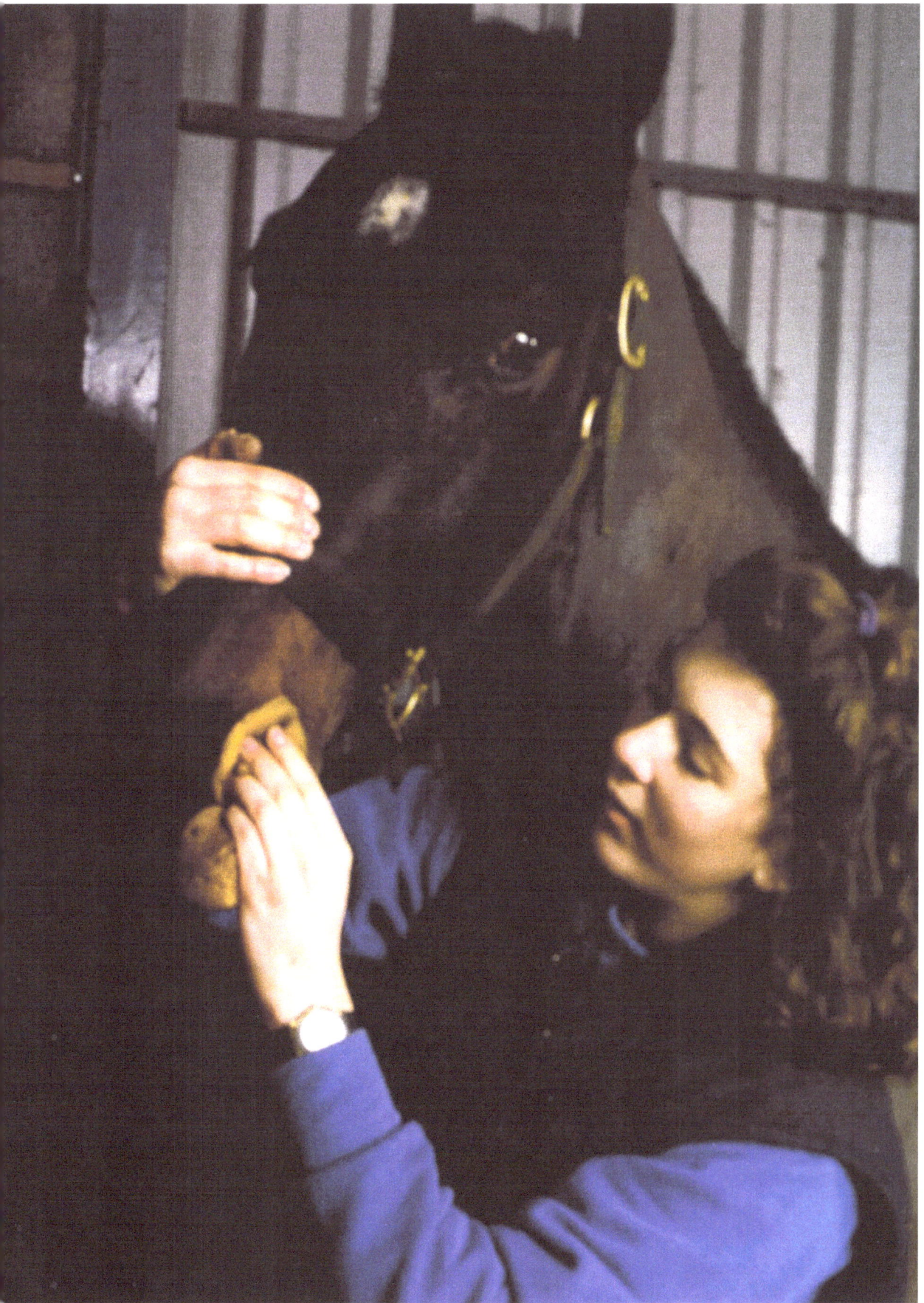

horses, watch instructional videos, and read any horse-related articles you can get your hands on. The more you learn about horses, the better off you'll be.

Eventually you'll be ready to take an equine sports massage therapy certification course. The course will teach you everything you need to know about how to massage a horse. You'll learn equine anatomy and physiology, how muscles work, how to recognize specific problem areas and injuries, and how to apply special massage strokes where they're needed most. Most of all, you'll learn more than you ever thought possible about horses. By the time you're through, you'll understand horses inside and out.

Outlook

Equine sports massage therapy is a relatively new field and is just now gaining acceptance in the horse world. There aren't many trained equine sports massage therapists out there, so now is a great time to begin your career and get a jump on the competition!

How much you make as an equine sports massage therapist will depend on where you work and on the rates charged by other therapists in your area. Most equine sports massage therapists find they can charge anywhere from twenty-five to eighty dollars for a sixty- to ninety-minute massage.

More important than what you charge for your services will be how you market yourself. You have to get to know the local veterinarians, trainers, and horse owners and must show them you're a skilled professional. You have to be able to show how what you do is of benefit to their horses. Then, once you've convinced them to give your work a try, you have to put your talent to work. As the saying goes, the proof is in the pudding!

Profile

I am twenty-seven years old and have twenty years of experience riding and training horses and working in barns. Last year I decided to try something new—equine sports massage therapy. I completed a two-week certification program through a company called EquiTouch in Loveland, Colorado. The program gave me the skills to practice equine sports massage therapy on my own.

Massage therapy has put me on a whole new level with horses. All horses

An animal massage therapist works with a horse before a race.

appreciate having their muscles rubbed. When you start massaging horses, they sometimes start moving around and itching themselves and fidgeting, but when they realize that what you're doing feels really good, they settle down and show clear signs of comfort and pleasure. They push into you for more pressure, they let their heads drop, and they move around less. When I see this, I know that what I'm doing is helping.

Another way I know my massaging helps is from what the owners tell me. Usually they say that their horses perform better and feel better to ride after I give them a massage or a series of massages. The fact is, horses are work animals— they jump, race, chase cattle, do dressage, and even make tight turns around barrels. Caring for their muscles is key to keeping them fit and healthy.

—Jen, an equine massage therapist
from Fort Collins, Colorado

Animal Careers: A Hot Job Market

Animals, of course, are a major part of modern-day life. Cats, gerbils, birds, and fish are all kept as pets. The sound of barking dogs is common in almost any neighborhood. It should come as no surprise, then, that careers in animal care—everything from working as a veterinary assistant to pet sitting—are booming.

Still, just because the jobs are there doesn't mean they're easy to get. "You've got to be willing to get in there and work," says Jerilee Zezula, an associate professor of applied animal science and the coordinator of the small-animal care program at the Thompson School of Applied Sciences in New Hampshire. "And if you want to move up the ladder, you have to be able to think and reason." Competition is keen, especially for positions that demand experience. Although some jobs require little more than a high-school diploma and a willingness to work, others are offered only to those with proper training and certification from professional organizations. To make matters more difficult, pay is often low considering the amount of time and effort required, and working conditions can be noisy, dirty, and physically demanding. "That's the problem with animal jobs," says Zezula, who is also a veterinarian. "A lot of people want to work in the field because they love animals, but they have no idea what they're getting into."

So what does it take to succeed? Courses in animal science are helpful, as is biology. If you have the

continued

41

COOL CAREERS WITHOUT COLLEGE

opportunity, try volunteering or, even better, working as a paid apprentice. Then again, if science isn't your thing or hands-on animal work sounds too hairy, consider a job where you aren't actually handling the animals—like a receptionist at a clinic, for example.

Most of all, the key to a career in animal care, whether it be as a trainer of search-and-rescue dogs, as a breeder, or as a pet store employee, is remembering just what attracts you to the field in the first place—animals. If you can do that, you'll be happy no matter where you end up.

FOR MORE INFORMATION

BOOKS

Denoix, Jean-Marie. *Physical Therapy and Massage for the Horse: Biomechanics Exercise Treatment.* North Pomfret, VT: Trafalgar Square, 2001.

Hourdebaigt, Jean-Pierre. *Equine Massage: A Practical Guide.* New York: Howell Book House, 1997.

Scott, Mike. *The Basic Principles of Equine Massage/Muscle Therapy.* Boston: Massage/Muscle Therapy Productions, 1997.

Way, Robert F. *Anatomy of the Horse.* Philadelphia: Lippincott, 1965.

WEB SITES

About Horses
http://www.horses.about.com

Equine Holistic Health
http://equinehealth.alphalink.com.au

Equisearch
http://www.equisearch.com

Harmony Touch Equine Sports Massage Therapy
http://www.harmonytouch.com

Horse Click
http://www.horseclick.com

HorseWeb
http://www.horseweb.com

The Michigan Horse Page
http://www.horsepage.com

MAGAZINES

The American Quarter Horse Journal
P.O. Box 200
Amarillo, TX 79168
(806) 376-4811
Web site: http://www.aqha.com

Equus
P.O. Box 420235
Palm Coast, FL 32142-0235
(800) 829-5910

Horse & Rider
1597 Cole Boulevard, Suite 350
Golden, CO 80401
(303) 445-4700
Web site: http://www.horseandrider.com.

The Horse: Your Guide to Equine Health Care
The Blood Horse, Inc.
P.O. Box 4680
Lexington, KY 40544-4680
(606) 276-6771
Web site: http://www.thehorse.com

Practical Horseman
P.O. Box 420235
Palm Coast, FL 32142-0235
(877) 717-8929

The Western Horseman
Western Horseman, Inc.
P.O. Box 7980
Colorado Springs, CO 80933-7980
(719) 633-5524
Web site: http://www.westernhorseman.com

ASSOCIATIONS

International Association of Equine Sports Massage Therapists
P.O. Box 447
Round Hill, VA 20141
Web site: http://www.iaesmt.com.

OTHER INFORMATION

EquiTouch Systems
P.O. Box 7701
Loveland, CO 80537-0701
(800) 483-0577 (for certification as an equine sports massage therapist)

4

VETERINARY TECHNICIAN

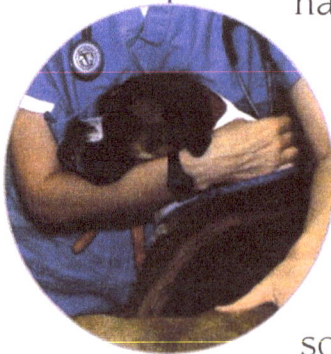

If you have a pet, chances are you've spent some time at your local veterinary hospital or clinic. You've seen the veterinarian at work, holding animals, preparing for surgery, doing tests and prescribing medicine, and perhaps you've thought it would be fun to work as a vet someday. But did you know that you don't necessarily have to be a veterinarian to work in a vet hospital? Today most veterinarians rely on help from highly trained assistants known as veterinary technicians. Vet techs, as they're called, play an incredibly important role in many aspects of animal care and treatment.

Description

Veterinary technicians working in veterinary hospitals do so under the supervision of veterinarians and have many important responsibilities. They talk to and explain procedures to pet owners, record patient histories, and do lab work. They prepare instruments, equipment, and animals for surgery, and often assist the veterinarians with medical and surgical procedures. They treat animals for diseases and medical conditions, and do things like develop X rays, examine blood samples, and prepare tissue samples. They might splint a broken leg, wrap an open wound, or just help to calm or hold an animal while the veterinarian looks at it. They also might prepare vaccines for the prevention of common diseases. Another important part of the vet tech's job is cleaning and sterilizing instruments and materials after a routine exam or surgery is performed.

Education and Training

You can start preparing for a career as a veterinary technician right now while you're still in junior high school.

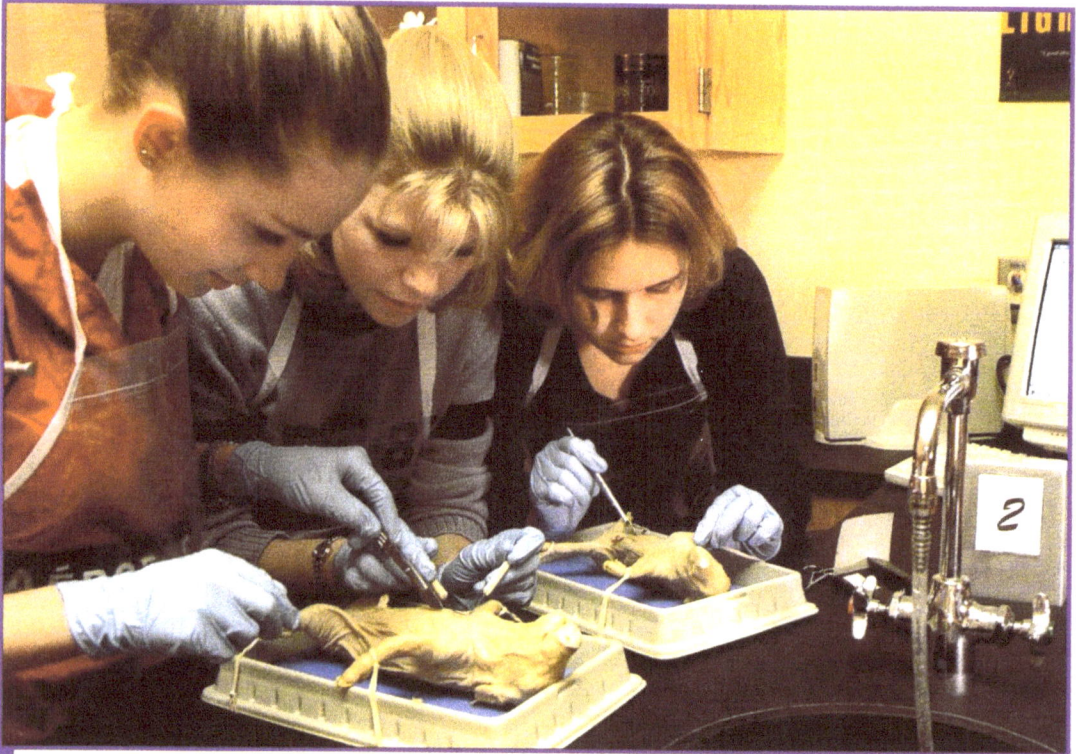

Biology students practice dissection.

Work hard in your science, math, and computer courses, and, once you start high school, try to take as many biology and animal science courses as you can.

Many states require veterinary technicians to graduate from a special two-year training program approved by the American Veterinary Medical Association. The certification programs, which are offered throughout the country, teach future vet techs about animal care, nutrition, and anatomy and physiology. They include challenging courses in math, chemistry, biology, and other subjects. They also teach

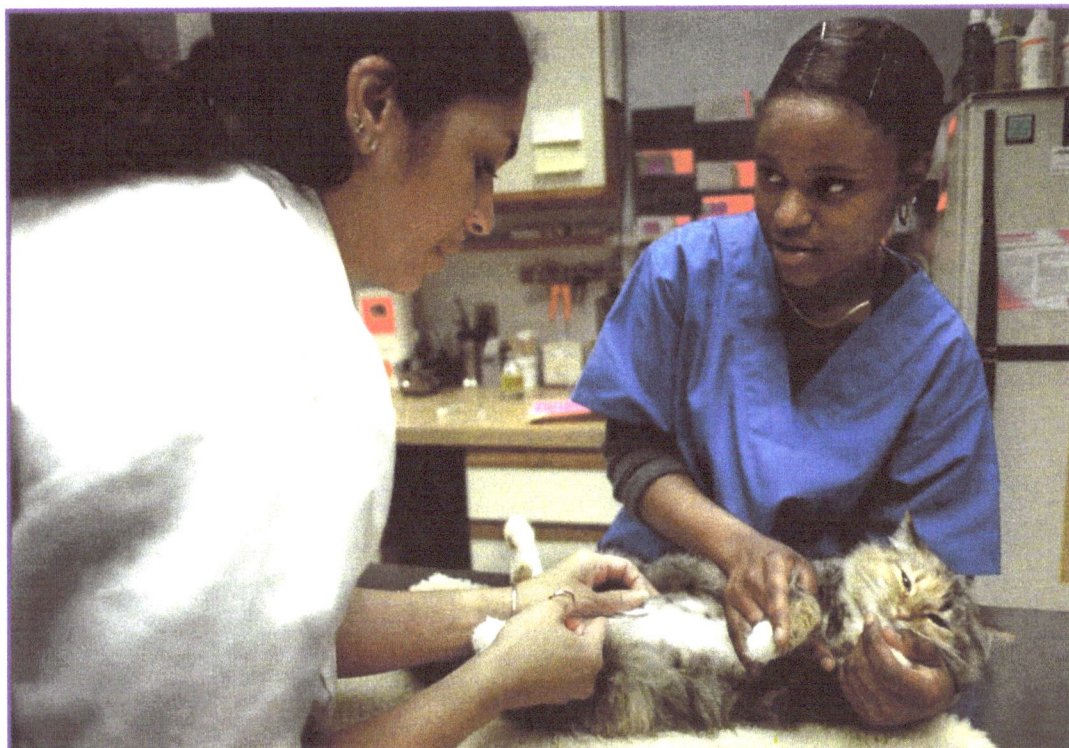

A high-school student assists a vet as she removes stitches from a cat after surgery. The student works at the vet's office as part of an internship to learn about career possibilities.

practical vet-tech skills like how to use medical instruments, how to administer anesthesia, and how to assist a veterinarian during surgery. They use real live animals, so students get hands-on experience. During high school, try to get experience working with animals by volunteering at a local vet hospital or even a kennel or an animal shelter. Then, when you graduate with your diploma, you'll be qualified to enroll in a vet-tech program.

When you finish the vet-tech program, you'll be eligible to take the state veterinary technician exam in whatever state you want to work. In many states you must pass this exam to get a job.

Outlook

The U.S. Bureau of Labor Statistics reports that in 1998 there were 32,000 professional veterinary technicians. They also report that this number is likely to grow in the next ten years, as more and more employment opportunities for vet techs are becoming available.

While most vet techs today work at private veterinary hospitals, in the future there will be more jobs elsewhere. Zoos need vet techs to care for their animals. Universities hire vet techs to teach and to help conduct research. Humane societies and animal control operations need vet techs, too.

How much you make as a veterinary technician depends on where you work, how much experience you have, and what kinds of responsibilities you're given. Expect to earn anywhere from $15,000 when you first start out to $40,000 as you gain more experience.

Profile

I am twenty-five years old, and have been fascinated with animals since I was very little. My ambition when I was young was to one day become a veterinarian. I was fortunate to grow up in a home where animals were considered part of the family. I don't think there was a period in my life in which I didn't have a cat or a dog, sometimes even four at once! Right now my career choice is to be a veterinary technician. It is not the same as being a veterinarian, but it allows me to work with animals and to work with patients that I love.

When I go to work, I never know what to expect. Over the years, I have encountered many different animals and situations. Some days animals will come in who just need their vaccinations. Other days, there have been animals that come because they've been hit by a car or they've been shot. It's amazing the ability that these animals have to heal and to endure. I have witnessed an animal who came in who had been hit by a car and had her leg hanging on only by skin. When she was on the table being examined, I patted her tummy and she just wagged her tail.

Veterinary technicians treat a dog in the emergency room at the Ohio State University Veterinary Hospital in Columbus, Ohio.

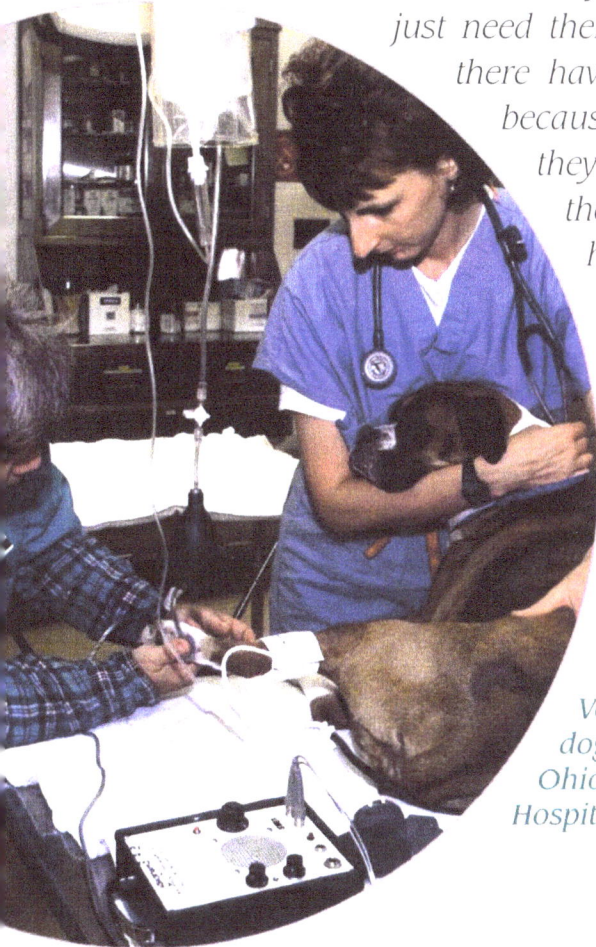

Even though she was in pain, the touch of my hand on her belly was enough to make her happy.

Another part of my job involves dealing with more difficult situations. When you work in a veterinary clinic, you get to know your patients. You feel a bond with them and their owners. There is a time, though, when the pet may get sick or may get hit by a car and the doctors cannot do anymore to save it. Then the owner has to decide whether euthanasia, or putting the pet to sleep, is best for the animal. That is when my job is the hardest.

Being a veterinary technician involves a lot of grunt work. My job lets me handle cute, cuddly creatures, but it also has me handling creatures that bite and scratch. It can, like any career with animals, be dangerous at times. I have been bitten by a dog and scratched and bitten by a cat. Both times I had to go to a doctor for my wounds.

If you think you might want to be a veterinary technician, take a day and visit a veterinary hospital to see what it's like. And ask yourself some simple questions like: Can I handle animals in pain? Does blood bother me? Do weird smells make me sick? These are situations that you will have to deal with if you work as a veterinary technician.

—Pilar, a vet tech from
Santa Fe, New Mexico

Making Cents Out of Summer

Summer, some say, is the sweetest of seasons. And with long days, lots of sunshine, and no school, who would argue otherwise? But if you're like many students, summertime spells more than countless hours spent soaking in the rays or hanging out with friends. Indeed, unless you've found a way to make money grow on trees, your three-month "vacation" likely includes a four-letter word: work.

Holding down a job, of course, doesn't mean the end of summer fun. In fact, a little hard work is a good idea for several reasons. For one, you'll get your first taste of what it's like to be financially independent. You can use the cash you earn to pay off bills or buy things you "need," like movie tickets, clothing, or a bike. Second, if you play your cards right you might find your work experience to be the perfect stepping stone toward a better job the following summer or even toward a future career.

If you've already bagged the job of your dreams, you're a step ahead of the game. If not, it's time to get in gear. "There is a job out there to fit almost any personality," says Marshall Brain, author of *The Teenager's*

continued

Guide to the Real World. "The important thing to recognize is you do have options." In determining just what those options are, it's a good idea to assess your situation and the reasons you need a job in the first place. Do you need to make $3,000 because that's the only way you'll be able to pay all of your expenses? Or is money not an issue? If how much money you pocket is not high on your list of priorities, consider an unpaid internship with a local business or volunteering somewhere like an animal hospital. "You won't make any money," says Brain, "but the experience can be invaluable."

FOR MORE INFORMATION

WEB SITES

American Society for the Prevention of Cruelty to Animals
http://www.aspca.org

Vetlearn.com
http://www.vetlearn.com

Vet Tech.com
http://www.vettech.com/weblinks.htm

MAGAZINES

American Journal of Veterinary Research
American Veterinary Medical Association
1931 North Meacham Road, Suite 100
Schaumburg, IL 60173-4360
(800) 248-2862

Veterinary Technician Magazine
275 Phillips Boulevard
Ewing, NJ 08618
(800) 426-9119

ASSOCIATIONS

American Veterinary Medical Association
1931 North Meacham Road, Suite 100
Schaumburg, IL 60173-4360
(847) 925-8070
(800) 248-2862
Web site: http://www.avma.org.

Canadian Veterinary Medical Association
339 Booth Street
Ottawa, Ontario K1R 7K1
Canada
(613) 236-1162
Web site: http://www.cvma-acmv.org.

North American Veterinary Technician Association
P.O. Box 224
Battle Ground, IN 47920
Web site: http://www.avma.org/navta

5

PET GROOMER

In the world of animal care, the pet groomer is to cats and dogs what the beautician is to people. Pet groomers make pets look good. They untangle knotted fur. They cut hair to just the right shape and length. They trim nails. They even bathe animals and blow-dry them. Whatever it takes to make a pet look its best, the groomer does it. Got a knack for style? This may be your calling.

Description

Groomers can work in many different places. Some work in kennels. Others work in veterinary clinics or in animal shelters. Still others find employment in pet supply stores or pet shops. One thing all of these groomers have in common is their concern for appearance. Their job is to make the pets in their care look perfect—or at least as perfect as possible.

Many groomers are self-employed. They run their own grooming businesses, often hiring assistants to help them with the workload. In addition to their regular grooming duties, they must tend to business concerns like answering telephones, scheduling appointments, and paying bills. They also must keep track of their expenses, pay their employees, and keep detailed records of the pets they've seen and what was done to them in the grooming process.

The actual grooming process, when the pet is worked on by the groomer, is typically straightforward. The pet is put on a grooming table and brushed to remove any twigs, grass, or big clumps of dirt or dust. Then the groomer uses electric clippers and special grooming shears to cut the hair and a sturdy comb to untangle and pull the hair into shape. Once the hair is cut to the groomer's satisfaction, he or she then trims the animal's nails, cleans its ears, and washes it to remove fleas, ticks, and other bugs. The wash is followed by a thorough towel or blow-dry and a final clipping and styling. Some groomers will then add special powders and perfumes to make the animal smell nice. When the groomer is finally through, the pet may look nothing like it did when it walked in!

Of course, it's not always that easy. Groomers have to be very good with animals. They must be able to calm them down so they're not nervous or likely to bite. If they can earn the trust of the animal, they'll have much better luck cleaning it up. If the animal doesn't cooperate, there are other options. Sometimes a veterinarian will give the animal medication to calm it down before it goes to the groomer, or the groomer will use a muzzle to cover the pet's mouth for protection from biting.

Education and Training

As with most jobs where you're around other people's pets day in and day out, you'd better love animals if you want to be a groomer. The dogs and cats you see will often be dirty, stinky, and shedding hair when they're placed on your table. You have to be able to see through the mess and imagine—and eventually create—perfection.

Start by spending as much time as you can around pets—your own, your neighbors', your friends'. Offer to

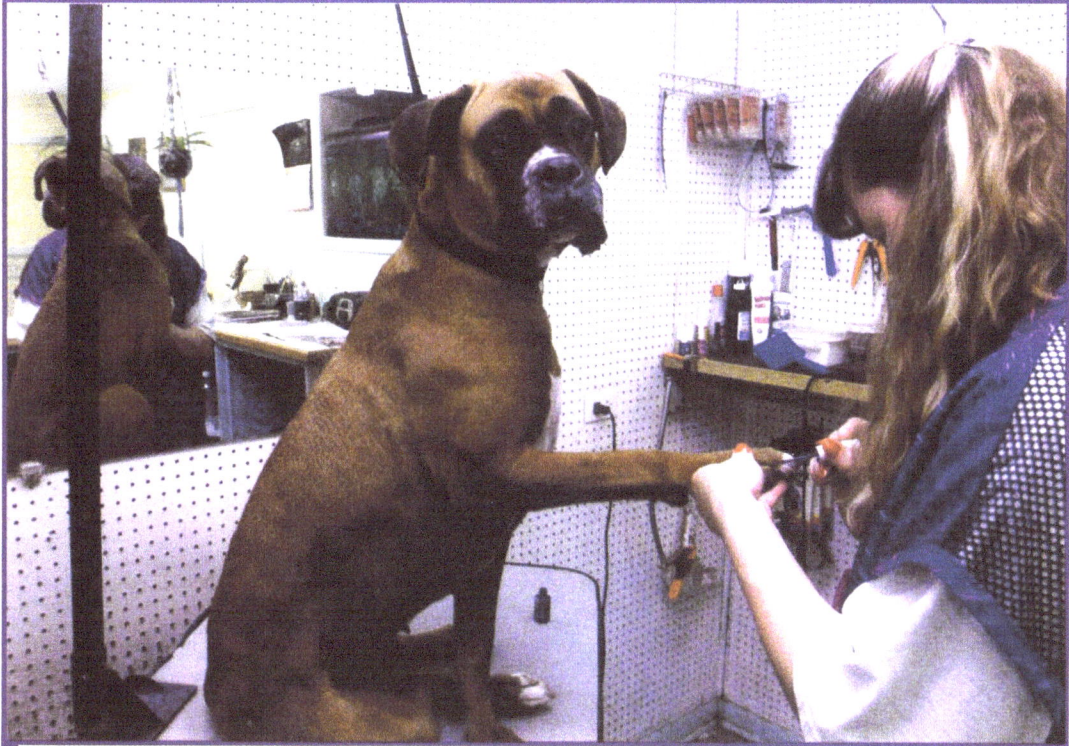

Full-service pet groomers perform services ranging from bathing pets to cleaning their ears and nails.

clean your friends' pets for free, and see how hard it is to keep an animal calm while you do strange things to it like comb its hair and wash its tail.

In high school, take as many courses as you can in subjects like business, health, math, and science. An important part of being a career groomer is knowing how to run your own business. It's very helpful to have good bookkeeping and office-management skills. Also take art and design classes, as creativity is a major part of being a good groomer. You need a great imagination to turn dirty creatures into works of art of

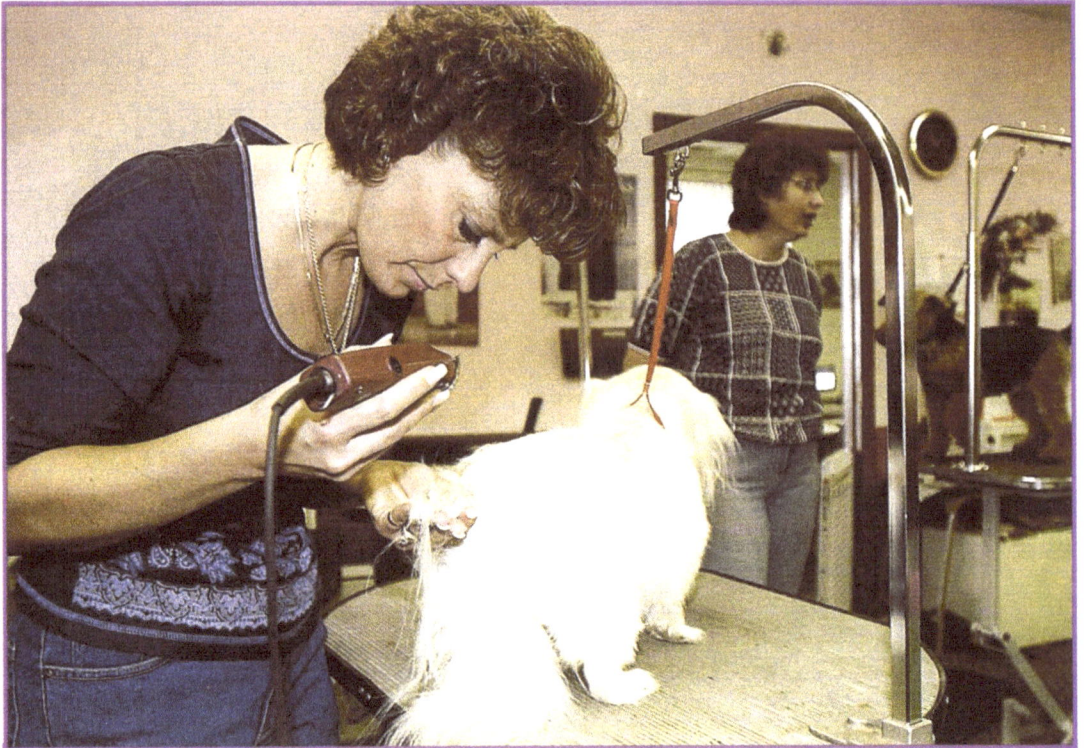

A pet stylist grooms a Pekingese for a show.

which their owners will be proud. Finally, read all the books you can find on grooming and pet care.

For now there are no certification requirements for groomers, but that may change in the future. Still, it's a good idea to take pet-grooming courses or to attend a pet-grooming school approved by the National Dog Groomers Association of America. Grooming schools train professional pet groomers. They're a great way to gain hands-on grooming experience. You'll learn bathing, brushing, and clipping techniques, as well as animal anatomy and physiology. You'll also learn proper styling techniques for different breeds—an important skill if you end up working with show dogs. Finally, you'll learn what it takes to run your own pet-grooming business. Upon graduation from such a school, you're qualified to take an exam. Pass the exam and you'll be NDGAA-certified—a great addition to your grooming résumé!

Most aspiring groomers also gain confidence in their skills by apprenticing with professionals. By working under an experienced groomer, or in any aspect of pet care, whether in a veterinary office, a pet shop, or a kennel, you'll learn everything you need to know and will be thoroughly prepared to begin a promising career.

Outlook

As long as people care about their pets, there will be a high demand for groomers. And more and more people are buying pets—especially dogs—so business is bound to grow. One interesting factor in the grooming field is how the economy influences business. When the economy is good, and people feel like they have plenty of money, they tend to pamper their pets. They think nothing of bringing their pets in for a quick trim, brush, or wash. When the economy is bad, however, business may slow down, as pet owners decide they'd rather hold onto their money than spend it on their pooches. Either way, don't expect to

A worker at an animal shelter washes a kitten in a sink.

make much more than $15,000 per year when you start out, or $25,000 per year once you're established.

In any case, if you decide to enter the grooming business, whether you succeed or not will largely depend on how motivated you are. If you become a top-notch groomer, build a name for yourself in your area, and work hard to gain new clients, you're almost guaranteed to do well.

A Cautionary Note

Dr. Jerilee Zezula of the Thompson School of Applied Sciences small-animal care program in New Hampshire, says:

> There are tons of jobs out there for groomers, but grooming is a very demanding job. It requires a lot of physical effort and energy. It's a good living, but it's hard work. Job burnout is high.

FOR MORE INFORMATION

BOOKS

Francais, Isabelle, and Richard Davis. *All (160) Breed Dog Grooming.* Neptune City, NJ: TFH Publications, 1988.

Gold, Charlotte. *Grooming Dogs for Profit.* New York: Howell Book House, 1986.

Pinney, Chris C. *Guide to Home Pet Grooming.* Hauppauge, NY: Barrons Educational Series, 1990.

Stone, Ben, and Pearl Stone. *The Stone Guide to Dog Grooming for All Breeds.* New York: Howell Book House, 1981.

Walin, Dorothy. *The Art and Business of Professional Grooming.* Loveland, Co: Alpine Publications, 1986.

WEB SITES

Groom & Board
http://www.groomandboard.com

Groomers Network.com
http://www.groomersnetwork.com

Intergroom 2001
http://www.intergroom.com

Online Grooming Board
http://www.dogpro.com/designs/groombbs/groomboard.html

Pet Business
http://www.petbusiness.com

PetGroomer.com
http://www.petgroomer.com

The Pet Groomer's Lounge
http://www.groomers.net

PetSytlist
http://www.petstylist.com

MAGAZINES

AKC Gazette
American Kennel Club
260 Madison Avenue
New York, NY 10016
(212) 696-8333
Web site: http://www.akc.org/akc/

Cat Fancy
Fancy Publications, Inc.
P.O. Box 57900
Los Angeles, CA 90057
(213) 385-2222
Web site: http://www.animalnetwork.com/cats/default.asp

Dog Fancy
Fancy Publications, Inc.
P.O. Box 57900
Los Angeles, CA 90057
(213) 385-2222
Web site: http://www.animalnetwork.com/dogs/default.asp

Groom & Board
H. H. Backer & Associates
200 South Michigan Avenue, Suite 840
Chicago, IL 60604
(312) 663-4040
http://www.groomandboard.com.

Groomers Voice
National Dog Groomers Association of America
Box 101
Clark, PA 16113
(742) 962-2711
Web site: http://www.nauticom.net/www/ndga

Groomer to Groomer
Barkleigh Productions, Inc.
6 State Road, Suite 113
Mechanicsburg, PA 17055
(717) 691-3388

Pet Business
McFadden Pet Business
233 Park Avenue South, 6th Floor
New York, NY 10003
(212) 788-7054
Web site: http://www.petbusiness.com

ASSOCIATIONS

International Professional Groomers
79 Flint Locke Drive
Duxbury, MA 02332

National Dog Groomers Association of America
Box 101
Clark, PA 16113
(742) 962-2711
http://www.nauticom.net/www/ndga

6

ANIMAL SHELTER EMPLOYEE

Animal shelters are also known as humane shelters. They protect animals and work for animal welfare. Stray dogs and cats often end up in animal shelters, as do abused or injured pets. Animal shelter employees are involved in the day-to-day business of running the shelter. They take care of every-thing, from feeding and caring for the animals to finding new homes for

An animal shelter employee works with a nervous cat.

them in the local area. It's not an easy job—animals are often very sick or may die from their injuries. Some are never adopted and must be put to sleep. Still, life as an animal shelter worker does have its rewards; the first time you nurse an animal back to health and see it off to a new and loving home, you'll see why.

Description

Most animal shelters are nonprofit organizations. That means they don't make any money over what they need to cover expenses—just enough to pay employees' salaries and keep things up and running. So if you decide to work in a shelter, don't expect a fat paycheck. Starting salaries often offer no more than $10,000 per year. You will be paid, of course, but you'd better be there because you love working with animals, not for the money.

Animal shelter employees hold many different positions. Animal caretakers, for instance, feed the animals, clean cages, and provide water. They also keep records of all the animals that are checked into the shelter and note when they are finally released to new homes. They keep track of all tests that are conducted at the shelter and medications and treatments that are given there, and sometimes they even do vaccinations if a veterinarian or veterinary technician is there to help. Finally, they often spend time working with the public, answering questions about humane pet care, how the shelter works, and which animals are available for adoption. Unfortunately, another job of the animal caretakers is euthanization. When they euthanize animals, they give them a special drug that painlessly puts them to death. This is done only as a last resort, when the animal is seriously ill or injured or when it's impossible to find a new home for it. This can be very stressful for the animal shelter employee.

Most of the animals at animal shelters are domesticated, like cats and dogs. Some shelters, however, also care for

The manager of an animal shelter in Savannah, Missouri, adjusts sprinklers and tarps to keep pens cool during the summer.

farm animals, and others specialize in caring for wild animals. Animal shelters that care for wild animals are usually called wildlife refuges or wildlife rehabilitation centers.

Other people who typically find work in animal shelters include humane investigators, humane educators, and shelter managers. Humane investigators do detective work when they hear about an abandoned or injured animal and see if they can find out how the animal ended up that way. If they feel the law was broken, they call the police and try to make sure whoever abused the animal is brought to justice.

They never lose focus on the animal, however, and bring those that need help to the shelter so they can be cared for.

Humane educators teach the public about humane treatment of animals. They travel around the local community and talk to people about what it means to be a responsible pet owner. The job of the shelter manager is to make sure the shelter runs smoothly and everything works the way it is supposed to work.

Education and Training

No special training is required of animal shelter employees, but helpful courses are offered through the Humane Society of the United States, the American Humane Association, and the National Animal Control Association. These organizations teach things like how to deal with a pet owner that has abused his or her pet, how and when to euthanize, and how to work with wild animals.

Practical experience is helpful when you start applying for animal shelter jobs. If there's a shelter near you, volunteer to help out after school or on the weekends. Or see if you can land a summer position, paid or unpaid. The best training for animal shelter employees is on the job, so do your best to get your foot in the door and you'll be off and running. As with most jobs in the animal world, as you gain more experience you'll have opportunities to move up the ladder into management positions. Eventually you just might find yourself running the show!

Outlook

For the foreseeable future, at least, there should be no shortage of job openings in animal shelters. The fact is, many people get pets and don't know how to care for them. Often these pets end up neglected, abused, or on the street. Without animal shelters, they'd have nowhere to go. Fortunately, more and more animal shelters are being established around the country to meet the growing demand for humane animal care. If this is your field, you can certainly count on plenty of work. Nancy Peterson of the Humane Society of the United States says:

> It's one thing to really love animals and want to work with them, but you have to realize that they're not always going to be the same as the pet you might have at home. Sometimes it can be a very sad experience to work with animals. There are a lot of animals and not enough people to help them. If you want to work with animals you also need to like working with people. Often there's a person attached to that animal. You need good people skills.

FOR MORE INFORMATION

WEB SITES

The Helping Paw Society
http://www.helpingpaw.org/shelters.html
A worldwide directory of animal shelters

ASSOCIATIONS

The American Humane Association
63 Inverness Drive East
Englewood, CO 80112-5117
(303) 792-9900
(800) 227-4645
Web site: http://www.americanhumane.org

The American Society for the Prevention of Cruelty to Animals (ASPCA)
424 East 92nd Street
New York, NY 10128-6804
(212) 876-7700
Web site: http://www.aspca.org

The Humane Society of the United States
2100 L Street NW
Washington, DC 20037
(202) 452-1100
Web site: http://www.hsus.org

People for the Ethical Treatment of Animals (PETA)
501 Front Street
Norfolk, VA 23510
(757) 622-PETA (7382)
Web site: http://www.peta-online.org

The World Society for the Protection of Animals
34 Deloss Street
Framington, MA 01702
(508) 879-8350
Web site: http//www.wspa-americas.org

7

PET SITTER

In many ways, pets are like babies. They usually can't feed themselves. They can't let themselves outside to go to the bathroom. They can't be left alone for long periods of time. For that reason, it's no surprise that professional pet sitters are in high demand. Pet sitters care for pets when their owners are away. Does going to a pet's home once, twice, even three times a day to feed, walk, and play with it sound like fun? Do you want to run your own business? If so, this could be your ideal career.

Description

Pet sitters are usually self-employed. They advertise in local papers and on billboards, set up schedules, make phone calls, and do all the grunt work required to run any small business. It's hard work, but it's also rewarding.

Although it might sound like fun going from house to house to check in on your favorite animals, pet sitting is not always glamorous. Pets get sick. They go to the bathroom on the floor. Sometimes dogs try to fight with other dogs when you take them for a walk. You might have to walk a dog in the middle of winter, in the freezing cold. Pets will lick you, get dirt on you, and shed their hair on you. If you don't like to get dirty, don't even think about pet sitting.

As a pet sitter you'll often spend time taking care of the pet owners' houses as well as their pets. For instance, they might ask you to pick up their newspapers in the morning, retrieve their mail during the day, turn on their lights at night, and water their plants. They might even pay you to spend the night to make absolutely sure all of the pets as well as the home are safe and sound. If an owner is gone for a long time—a month or more is often the case—you might begin to feel like you actually live in his or her house.

As far as the actual pet sitting goes, you'll have to be good at following the owners' directions. Usually they'll leave you notes describing everything they want you to do, or you'll talk with them before they leave to go over all the details. You'll have a list of phone numbers you'll be expected to call in an emergency—the vet, friends and neighbors, family, even the plumber, electrician, or housekeeper. Typical duties might include feeding, watering, walking and exercising, cleaning cages and kitty litter boxes, and giving medications to animals that need them. You'll be a pet's companion while the owner is away. You'll be responsible for the pet from the moment the owner walks out the door to the moment he or she returns.

Education and Training

Here's good news: You can start pet sitting right now, before you even get out of school! All you need to do is ask your neighbors if you can watch their pets while they're away. You can do it for free until you gain experience, or you can charge for your services right from the get-go. You'll be surprised how quickly you'll find work.

Another great way to prepare for a pet-sitting career is by finding work in any field that involves animal care.

A professional dog walker often handles several dogs at once.

With experience working for a veterinarian, a pet shop, a groomer, or any other pet care operation, you'll learn what it takes to work with animals as well as gain connections in the pet care world.

To make real money at pet sitting, you'll need to do it full-time. And to work full-time you should have at least a high-school diploma. It's difficult running your own business, and a solid education will help you to succeed. Do well in your math, business, and science courses and you'll have a good foundation for starting out on your own.

The keys to a successful pet-sitting operation are hard work and good references. Once you work for a few customers and show them how well you do your job, they'll pass around your name to friends. Before you know it, you'll have more work than you can handle.

If you decide you want to take it to the next level, you may get certified through Pet Sitters International or the National Association of Professional Pet Sitters. PSI and NAPPS offer courses in animal health and care, how to run a business, and other subjects. The courses can be taken at home by reading books and watching educational videos. When you're through studying, you can take a test to prove you learned the material. If you pass, you'll be certified. Although certification is strictly voluntary, it gives you credibility. Pet owners see that you're certified and know you're a professional. And that, after all, is what running a business is all about.

Outlook

According to the National Association of Professional Pet Sitters, more than 60 percent of homes in the United States have pets. That means there are more than 100 million cats and dogs, as well as countless other pets like fish, gerbils, birds, and hamsters. It should come as no surprise, then, that in many parts of the country, especially urban areas, pet sitters are in high demand. People are busy. They have

full-time jobs, work long or odd hours, and like to split town and go on vacation. Usually their pets must be left behind. And that's where pet sitters come into play.

Pet owners recognize the excellent services offered by professional pet sitters. And they'd much rather leave their pets at home, in a familiar environment, than at a boarding kennel where they're liable to be uncomfortable and frightened. For pet sitters, this means business. How much an individual pet sitter can earn depends on where he or she is working and what types of animals he or she works with, but standard rates generally range from $8 to $15 per pet per visit. And with numbers like that, it's easy to see how things can add up—and fast.

Still, don't expect to make a lot of money when you first start out. Running your own business can be a challenge, especially come tax time. Pet sitters make anywhere from $5,000 to $50,000 or more per year. What you make depends on your business. Run it like a pro and you won't be disappointed. A pet sitter from San Francisco had this to say:

I like my job as a pet sitter. There is often a very strange dynamic that exists between owner and pet

A pet sitter gives a dog a bath.

which the owner seeks to replicate with the pet sitter. It is not unusual to receive novel-length instructions from owners complete with stage directions and dialogue. For example, "Does Brandy want some lovies?" might translate in one client's mind to, "Does Brandy want to eat processed cat food this morning or would she rather go outside and decimate the neighborhood bird population?" And on and on.

FOR MORE INFORMATION

BOOKS

Anderson, Karen. *The Cat Sitter's Handbook: A Personalized Guide for Your Pet's Caregiver.* Minocqua, WI: Willow Creek Press, 2001.

Doyle, Patricia A. *Sit and Grow Rich: Petsitting and Housesitting for Profit.* Dover, NH: Upstart Publishing Co., 1993.

Hill, Dan. *The Pet First-Aid Book.* New York: McGraw-Hill, 1986.

Mangold, Lori, and Scott Mangold. *The Professional Pet Sitter: Your Guide to Starting and Operating a Successful Service.* Portland, OR: Kitty and Home Services, 1991.

Moran, Patti J. *Pet Sitting for Profit: A Complete Manual For Professional Success.* New York: Howell Book House, 1997.

Price, M. J. *The Pet Sitter's Diary, and the Jam-Packed Pet Information Manual.* Zephyrhills, FL: M.J. Price, 1999.

Roth, Suzanne M. *The Reality of Professional Pet Sitting.* Princeton, NJ: Xlibris Corporation, 1998.

MAGAZINES

The World of Professional Pet Sitting
Pet Sitters International
201 East King Street
King, NC 27021-9161
(336) 983-9222
Web site: http://www.petsit.com

Available to Pet Sitters International members only; includes information and advice and covers topics like animal health, pet care, marketing, and managing a pet-sitting business.

VIDEOS

Pet Sitting: Getting Started
National Association of Professional Pet Sitters
6 State Road, Suite 113
Mechanicsburg, PA 17050
(717) 691-5565
e-mail: NAPPSmail@aol.com
Web site: http://www.petsitters.org

ASSOCIATIONS

National Association of Professional Pet Sitters
6 State Road, Suite 113
Mechanicsburg, PA 17005
(717) 691-5565
Web site: http://www.petsitters.org

Pet Sitters Associates, LLC
4014 Boardwalk Street
Eau Claire, WI 54701
(800) 872-2941 Ext. 25
Web site: http://www.petsitllc.com

Pet Sitters International
201 East King Street
King, NC 27021-9161
(336) 983-9222
Web site: http://www.petsit.com

8

PET SHOP WORKER

Lizards, snakes, parakeets, hamsters—you'll find these animals and more in any pet shop. You'll also find pet food, pet toys, kitty litter, dog pens, and just about everything and anything a pet owner could possibly want or need for his or her pet. So who runs the show? The answer is pet shop workers do. Pet shop workers include shop owners, managers, cashiers, and other employees. If you find yourself amazed by all the screeches, snorts, and smells of your local pet shop and don't mind a little sales work, you may want to give a career as a pet shop worker a try.

Description

Pet shops can be as small as a local, one-person operation or as gargantuan as a multinational, Internet-based company with hundreds of employees. Chances are you'll get your start somewhere in between—perhaps as a cashier working the register in a respected shop in your hometown.

As a pet shop employee, you'll be part of a team of people trying to sell pets and pet supplies to the public. You'll keep busy every day doing things like stocking shelves, taking inventory, ordering new products, helping customers, and cleaning up. You'll fill feed bins when they get low, place sale items on special racks, and set up signs and posters. You might have to sweep the floor and lock everything up at the end of the day.

Most pet shops sell animals as well as pet supplies. These stores need employees who will clean out cages and

Contented, well-cared-for puppies await new owners.

aquariums, feed animals on a regular basis, and walk and play with those animals that need exercise and human attention. The kinds of animals you'll end up working with will depend on the store, but they could include everything from cats and dogs to birds, fish, snakes, turtles, and rabbits. You might have to change the water in a fish tank, rearrange the rocks and sticks in a turtle's box, or feed a hungry snake live mice for dinner.

Many career pet shop workers end up running their own pet shop someday. It's only natural—if you spend enough time working in a pet store you end up learning the ropes so well you may wind up running the show yourself.

Education and Training

There is no specific training required for pet shop workers. Still, if you plan to run your own pet shop someday, it doesn't hurt to study business, mathematics, and science. You'll need good business skills to manage your store, especially if you plan to be successful.

If you're sure you want a career working in a pet shop, the best time to start is now. Go to your local pet store and tell them you're interested in learning all about it. Tell them you want to help out in any way they can use you. Who knows, they might hire you!

A young customer looks at the aquarium display at a local pet store.

If you can't get a paid position at a pet shop, try to volunteer. And if no volunteer jobs are available, look for work elsewhere in the pet care field to get experience. Once you have experience the job offers are sure to start coming in.

Specific skills that are necessary for pet shop workers depend on the position held, but good customer service skills and experience caring for animals are definite requirements, as is familiarity with pet toys, foods, and health. You'll be dealing with both people and animals on a daily basis, so get to know both.

The Job Hunt

Once you've determined your goals, it's time to pound the pavement. Before school lets out, consider asking your teachers whether they know of any job opportunities for kids like you. At home, try flipping through the phone book for promising businesses that may need help. Once you've compiled a respectable list, start making calls. Also, check out the classifieds section of your local newspaper. If you have access to a computer, jump online and give "networking" a shot. Sites like http://www.campjobs.com, http://www.coolworks.com, http://www.summerjobs.com, and http://www.quintcareers.com are great all-around resources that post thousands of jobs available nationwide. Finally, don't forget to keep an open mind as you look. Even if you don't find a job working with animals, you could get valuable work experience in a restaurant, a retail store, or an amusement park. Keep an open mind!

Outlook

The outlook for careers in the pet shop world is good. People are buying pets and pet products left and right, and as long as people love animals they'll continue to do so. If you're thinking about working in a pet store, you should have no trouble finding work.

Expect to be paid minimum wage when you first start out working in a pet store. Most stores are small and can't afford to pay their employees much money. Once you gain experience, however, you should get a raise. Eventually you'll earn enough to make a good living, especially if you decide to run your own business.

FOR MORE INFORMATION

WEB SITES

Allpets.com
http://www.allpets.com

Petco.com
http://www.petco.com

PetsMart.com
http://www.petsmart.com

Pets Warehouse.com
http://www.petswarehouse.com

MAGAZINES

Bird Times
Pet Business, Inc.
Web site: http://www.birdtimes.com

Cats and Kittens
Pet Business, Inc.
Web site: http://www.catsandkittens.com

Dog and Kennel
Pet Business, Inc.
Web site: http://www.dogandkennel.com

Pet Age
200 South Michigan Avenue, Suite 840
Chicago, IL 60604
(312) 663-4040
Web site: http://www.petage.com

Pet Business
McFadden Pet Business
233 Park Avenue South, Sixth Floor
New York, NY 10003
(212) 979-4800
Web site: http://www.petbusiness.com

Pet Product News
P.O. Box 57900
Los Angeles, CA 90057
(213) 385-2222
Web site:
http://www.animalnetwork.com/petindustry/ppn/default.asp

Pets International
Web site: http://www.pets.nl/
The international marketing magazine for the pet industry.

ASSOCIATIONS

American Pet Products Manufacturer Association
255 Glenville Road
Greenwich, CT 06831
(800) 452-1225
Web site: http://www.appma.org

Pet Industry Joint Advisory Council
1220 19th Street NW, Suite 400
Washington, DC 20036
(202) 452-1525

Pet Industry Joint Advisory Council Canada
2442 St. Joseph Boulevard, Suite 102
Orleans, ON K1C 1G1
(613) 834-2111
Web site: http://www.pijaccanada.com

World Wide Pet Supply Association
406 South First Avenue
Arcadia, CA 91006
(626) 447-2222
Web site: http://www.wwpsa.com

KENNEL WORKER

Many pet owners choose to board their pets at a kennel when they go away. Kennels provide basic care for pets, including a place to sleep, plenty of food and water, and regular exercise. Sometimes they also provide grooming services. As a kennel worker you'll be responsible for making sure that any pets left in your care remain safe, healthy, and happy until the owners return.

Description

Kennel workers spend most of their time caring for dogs and cats. When you first start out, your duties will include tasks like cleaning cages and dog runs, filling food and water bowls, and exercising the animals. You'll receive pets from their owners and will transfer them to their temporary cages. You may take the dogs on walks, encourage the cats to play with toys, or just sit by their cages and keep them company if they appear frightened or sad. Often the animals in kennels are very nervous. They're in an unfamiliar environment, their owners are nowhere to be found, and there are dozens of other noisy animals nearby. It's a high-stress situation for a pet that is used to having its own space and home. One of your jobs as a kennel worker will be to help the animals be comfortable.

As you gain more experience, you'll be given more responsibilities. Your boss might ask you to provide basic health care for the pets in the kennel. You'll have to bathe animals, trim nails, clean paws and ears, and take care of other simple grooming needs. You also may do a little sales work. Kennels often sell pet food and supplies to their customers and need employees who can offer good customer service.

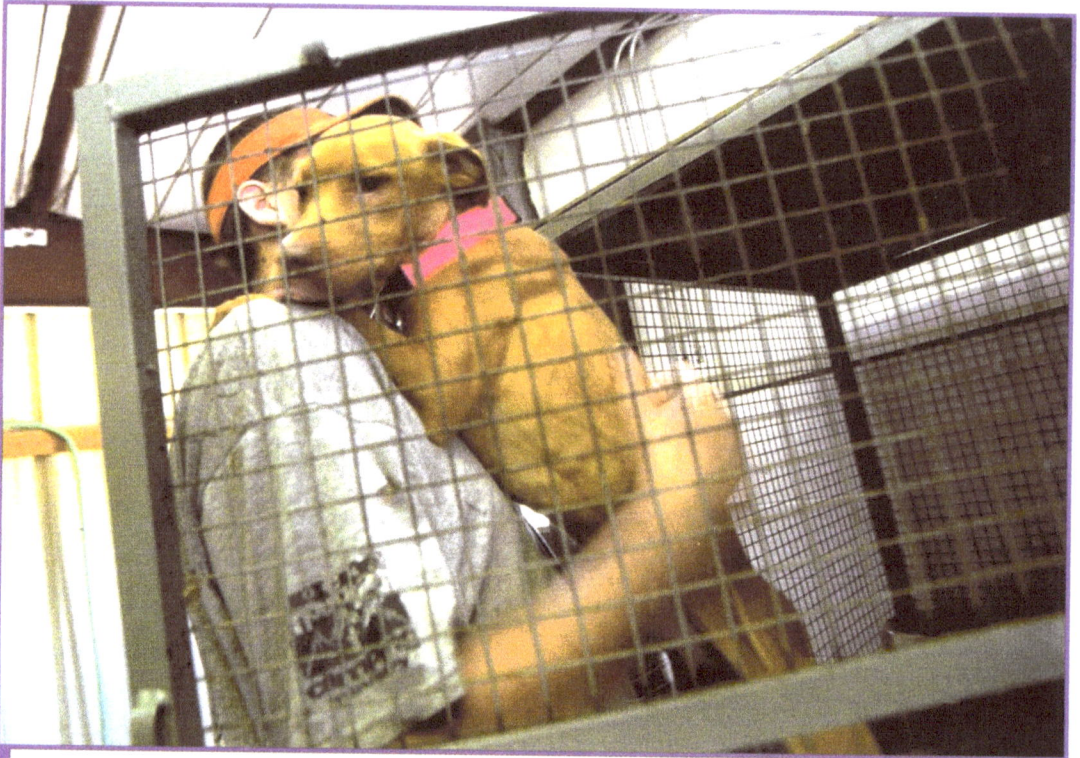

A kennel worker feeds and cares for animals.

Eventually, once you learn everything there is to know about being a basic kennel worker and have acquired plenty of experience, you might be promoted to a position as kennel supervisor or manager. These are leadership positions that require a thorough understanding of how your particular kennel operates and how to care for all the different breeds of animals in your kennel. You'll supervise other employees and make sure they do their jobs effectively. You'll be responsible for hiring and firing, purchasing new supplies, and running the business when the owner is not around.

Even though you'll be working with animals day in and day out, it's important to have good people skills as a kennel worker, no matter what your position. You'll need to be able to ensure pet owners that their pets are safe with you and that you'll take good care of them. For many pet owners it's not an easy decision to leave their favorite companion with a stranger. You have to convince them their pets will be safe.

Education and Training

Most beginning kennel workers learn their skills on the job. By going to work and just following the directions of their supervisors, they learn what it takes to be an effective and efficient animal caretaker. If this is the route you want to take, you can get started right away. See if your local kennels are hiring part-time or weekend help, or if they need volunteers. It's a great way to get your foot in the door.

Once you have some kennel experience you may decide you want to work in kennels as a career. If that's the case you should consider taking classes offered by the American Boarding Kennels Association. The first course of study offered by the ABKA is the Pet Care Technician Program. Many kennel operators require their employees to complete this program, which teaches the basic principles of animal care, handling, kennel management, and customer relations.

This course can be taken at home by reading required materials and taking a test when you're finished.

The ABKA's Advanced Pet Care Technician Program is a higher-level course. It's offered to kennel workers with at least a year of experience in pet care. It covers many of the same subjects taught in the basic pet care program, but in more depth. It also includes details on how to manage employees and run a business—important skills if you think you might want to advance to a management position.

Once you've finished the Advanced Pet Care Technician Program, you can enroll in the Certified Kennel Operator Program. This is the most advanced program offered by the ABKA. This course is for people who have at least three years experience working in a boarding kennel and who want to become a kennel operator. As a certified kennel operator, or CKO, you'll have exceptional animal care and business skills and will be recognized as one of the leaders in your field.

Outlook

You'll always be able to find jobs in the boarding kennel industry, but you'll probably have to start out at the bottom

A kennel worker rubs shampoo into the wet fur of a kitten.

of the ladder. Most beginning kennel workers are paid minimum wage or slightly above minimum wage; those with experience earn more. Be patient while you learn the ropes and you'll have no trouble succeeding in a kennel-related career.

FOR MORE INFORMATION

BOOKS

Allen, Dana G. *Small Animal Medicine.* Philadelphia: Lippincott, 1991.

Caras, Roger. *Pet Medicine: Health Care and First Aid for All Household Pets.* New York: McGraw-Hill, 1977.

Johnston, Dudley E. *Exotic Animal Medicine in Practice.* Trenton, NJ: Veterinary Learning Systems, 1991.

Tully, Thomas N. *A Technician's Guide to Exotic Animal Care.* Lakewood, CO: AAHA Press, 2001.

MAGAZINES AND NEWSLETTERS

Boarderline Newsletter
1702 East Pikes Peak Avenue
Colorado Springs, CO 80909
(719) 667-1600
Web site: http://www.abka.com
For members of the ABKA

Pet Services Journal
1702 East Pikes Peak Avenue
Colorado Springs, CO 80909
(719) 667-1600
Web site: http://www.abka.com
For members of the ABKA

ASSOCIATIONS
American Boarding Kennels Association
1702 East Pikes Peak Avenue
Colorado Springs, CO 80909
(719) 667-1600
Web site: http://www.abka.com

10

WILDLIFE CONTROL AND RELOCATION SPECIALIST

Sometimes wildlife, including squirrels, raccoons, birds, and other wild animals, end up in places that are, well, definitely not wild. Take the hawk that accidentally crashes through a window and into a family's living room, or the unfortunate squirrel that decides to build its nest in a chimney. These animals do not belong in

WILDLIFE CONTROL AND RELOCATION SPECIALIST

people's homes. They must be removed and relocated to a safe and appropriate place. Wildlife control and relocation specialists do just that.

Description

This is not a job for the timid. You've got to be fearless—not afraid to get down on your belly and crawl into small spaces, grab a ladder and climb to the very top of a roof, or duck and poke through spider webs and darkness in search of a hole or nest. The work can be dangerous; wild animals often have teeth and claws and won't hesitate to bite if they're afraid or are trying to protect themselves or their young.

You also have to be strong and healthy. You'll be working in rain or sunshine, in the hot summer or the dead of winter, in icy cold basements and scorching attics. You'll be carrying cages and loading and unloading equipment from your truck. You'll be on your hands and knees peeking into cracks and holes and pulling yourself onto rooftops to inspect chimneys and vents.

But strength isn't all you'll need. You'll also have to be good with people, as your customers will probably be right there with you as you do your work. You'll need to know how to be patient and how to explain what you're doing when customers ask.

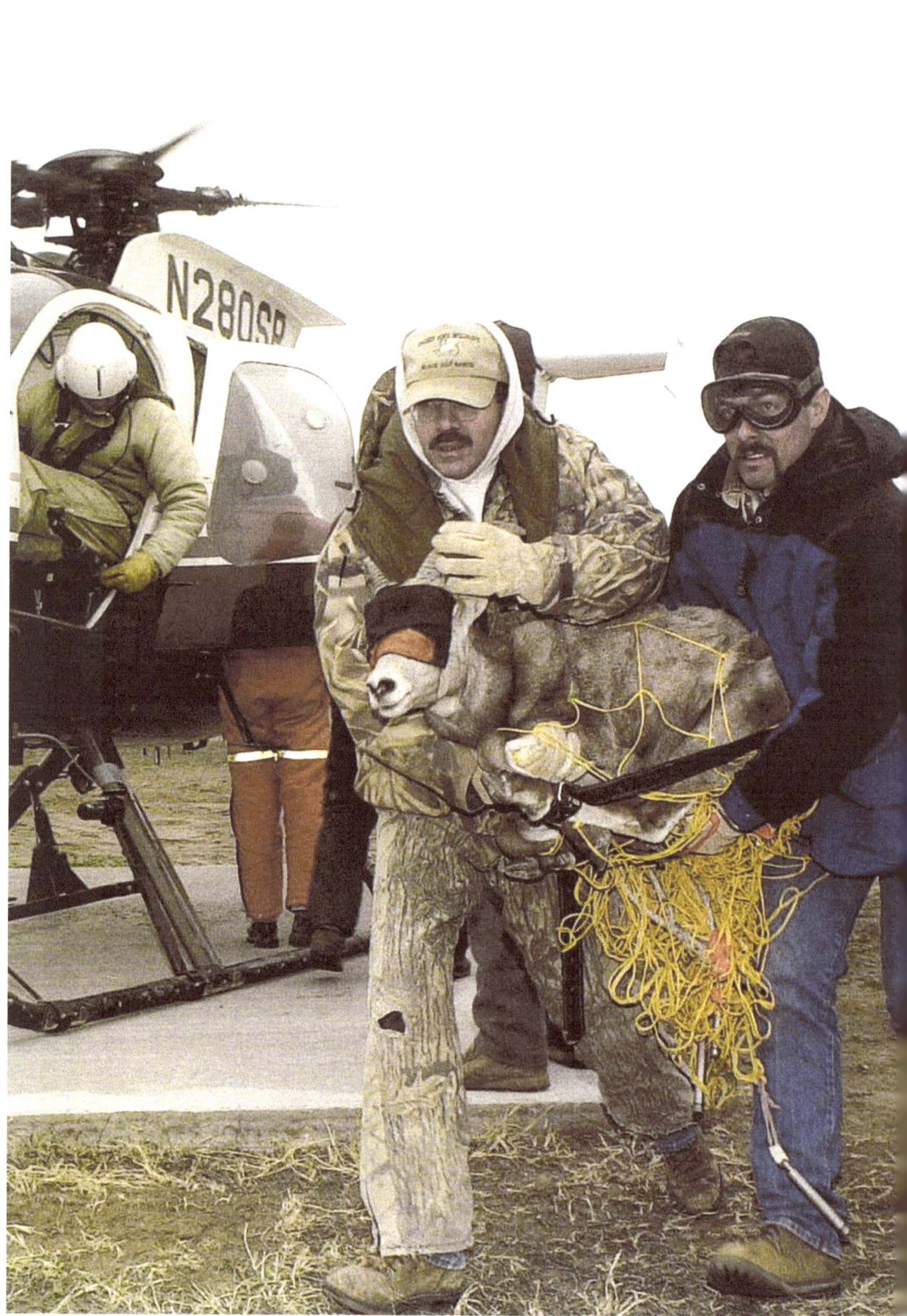

Wildlife control and relocation specialists have two major jobs. First, they remove problematic wildlife from urban areas—places where there are lots of people, homes, cars, and businesses. Second, they try to prevent wildlife from reentering those areas by relocating animals to their natural environment. The goal is to use humane techniques any time an animal is trapped or handled to avoid causing stress or injuries.

One way wildlife control and relocation workers prevent animals from entering homes is by placing sturdy steel screens over potential entryways like chimneys and vents. This prevents birds from flying down chimneys and into houses, and animals like squirrels and chipmunks from squeezing their way through air vents and into attics. Another way to discourage animals from reentering a home in which wildlife once lived is by deodorizing. Deodorizing removes the animal's scent. If the scent isn't removed, other animals will be tempted to come and investigate.

Once an animal is removed—usually with the use of harmless traps—it is relocated. The wildlife control and relocation professional drives the animal to a place where it is

Wildlife specialists relocate wild bighorn sheep to better pastures.

A wildlife worker sets a trap for beavers in an effort to control their population.

safe, can find food and shelter, and can live comfortably with other wildlife—like a park, for example. If the animal were relocated to a place where it wasn't comfortable, it would probably just try to enter another home or building.

Education and Training

Many employers in the field of wildlife control and relocation require their workers to receive special training, usually as an apprentice on the job. This is important to ensure the safety and welfare of both animals and people.

The National Wildlife Control Operators Association (NWCOA) offers a formal professional certification program for workers involved in wildlife control. Professionals with expertise in wildlife damage management and the resolution of wildlife conflict with humans and at least three year's experience may enroll in NWCOA's training courses and eventually be certified as wildlife control professionals. Those with less experience in the field may receive training for certification as apprentice wildlife control professionals.

Earning these certifications is a challenge. You must take courses in wildlife biology, animal handling, health and safety, and other subjects. You have to learn technical skills, public relations skills, and business skills. You have to take more courses throughout your career to maintain the certification. Certification is not required to work in the field of wildlife control, and you might learn everything you ever need to know on the job and through hands-on experience. But it does give you credibility, and when you've got competition from other professionals, that might be what separates you from the pack.

Outlook

As the U.S. population continues to grow, houses are being built closer and closer to the natural homes of wild animals—forests, rivers, prairies, and mountains. Animals

A park ranger scans a section of Central Park in New York City for an alligator that was reportedly spotted lurking near the water's edge.

like raccoons, rabbits, and even hawks and bears are increasingly showing up in populated areas. Wildlife control and relocation specialists are needed to protect these animals from danger as well as to protect people and property. There's no doubt that in the future the conflicts between humans and wildlife will continue to increase.

For both wildlife and homeowners, this is bad news, but for wildlife control and relocation professionals, it means there will never be a shortage of work. If you plan to enter the profession, how easily you find employment and how much you are paid will depend on where you live, the specialized training you receive, and how big a problem the animal-human conflict is in your area.

The NWCOA Code of Ethics

1. I affirm my strict adherence to all laws and regulations pertaining to wildlife damage management.
2. I ascribe to a professional code of conduct that embodies the traits of honesty, sincerity, and dedication.
3. I will show exceptionally high levels of concern and respect for people, property, and wildlife.
4. I will promote the understanding and appreciation of the numerous values of wildlife and scientific wildlife management, as well as an appreciation for the economic and health concerns of humans adversely affected by wildlife.
5. I will be sensitive to the various viewpoints on wildlife damage management.
6. I will provide expertise on managing wildlife damage to my clientele upon request, within the limits of my experience, ability, and legal authority.

7. I will promote competence and present an image worthy of the profession by supporting high standards of education, employment, and performance.

8. I will strive to broaden my knowledge, skills, and abilities to advance the practice of commercial wildlife damage management.

9. I will, in good faith, select new or time-proven methods for resolving wildlife damage conflicts and give due consideration to humaneness, selectivity, effectiveness, and practicality.

10. I will treat my competition and clientele in a courteous manner and in accordance with honorable business practices.

11. I will encourage, through word and through deed, all commercial wildlife control operators to adhere to this code and to participate in state associations of commercial wildlife control operators.

Profile

I have loved animals since I was a kid, and at a very young age I knew that whatever I did in life it would involve animals. I am now twenty-six and work as a wildlife control specialist in Denver, Colorado. Before I got this job I volunteered at a local wildlife rehabilitation center to gain experience handling exotic animals. I soon discovered that I had a great passion for helping rehabilitate injured wildlife and releasing them back into the wild. It was a wonderful feeling. I worked in rehabilitation for almost two years. I dealt with a wide variety of animals—raccoons, squirrels, foxes, beavers, rabbits, and some domesticated animals like pigs and ducks. Springtime was always very busy, with many baby animals being abandoned by mothers who had been injured and could no longer care for them. Our time was spent primarily bottle feeding baby raccoons and squirrels. It was always rewarding to watch them grow up healthy and strong, though it was also important to make sure they did not get attached to humans. Otherwise they would not be able to survive out in the wild on their own.

I eventually realized that I wanted to start a career in wildlife control and relocation. I soon found a company that was hiring and I decided to go for it. The company responds to homeowners and businesses that have problems with wildlife. (We don't deal with domestic animals like dogs or cats.) Whether it's squirrels in the attic or skunks living under a porch, we can help. I love my job and look forward to it every day. There is no typical day in this sort of work— you just never know what to expect.

One of my first jobs was to help a woman that had a raccoon living under her back porch. It would come out from under the porch and destroy her yard and garden at night. I jumped in my truck and headed over to her house to take a look around and set up traps. The traps we use are very humane and do not hurt the animals at all. I put marshmallows and honey in the traps. (Raccoons love anything sweet.) Since raccoons sleep during the day I was pretty sure it was under the deck while I was setting up my traps, so I blocked the traps up to the hole and left, hopeful that in the morning there would be a raccoon waiting for me.

Sure enough, the next morning the woman called to tell me we caught it! I drove to her house and found a very frustrated male raccoon, growling and making faces at me to show how upset it was that I had caught it. Before I left the job I closed up the area that the raccoon was getting into by

A wildlife specialist tags a baby peregrine falcon that nests in an urban environment.

patching the hole and burying wire under the ground around the porch to keep other animals from digging in. I loaded the raccoon into my truck and found a wonderful new home for him by a river, about twenty-five miles away from where I trapped him.

I use this same procedure when trapping squirrels out of attics or chimneys. Recently I had to rescue an American kestrel, a small bird of prey, out of a fireplace. I also work with bats, skunks, gophers, beavers, and any other small mammals that cause problems for people.

What I like most about this job is that I'm helping both people and animals by safely relocating the animals to a new home and relieving homeowners of their worries. I love the fact that I never know what to expect from day to day and I'm always sure to have some adventure along the way.

The most challenging aspect of this work is trying to understand the behaviors of different species of animals in order to figure out the best way to catch them. The job also requires climbing up on roofs to do inspections to try and figure out where the animals are getting into the houses. For someone who is afraid of heights, this may be a challenging thing to do.

I would definitely recommend this type of work to anyone who loves animals, enjoys being outdoors, and likes the idea of going on an adventure every day!

—Lisette of Golden, Colorado

COOL CAREERS WITHOUT COLLEGE

FOR MORE INFORMATION

BOOKS

Hadidian, John. *Wild Neighbors: The Humane Approach to Living with Wildlife.* Golden, CO: Fulcrum Publishing, 1997.

Landau, Diana. *Living with Wildlife: How to Enjoy, Cope with, and Protect North America's Wild Creatures Around Your Home and Theirs.* San Francisco: Sierra Club Books, 1994.

WEB SITES

AAA Wildlife Control
http://www.aaawildlife.com
A Canadian company

MAGAZINES

International Wildlife
National Wildlife Federation
11100 Wildlife Center Drive
Reston, VA 20190-5362
(703) 438-6000
(800) 822-9919
Web site: http://www.nationalwildlife.org/intlwild

NACA News
P.O. Box 480851
Kansas City, MO, 64148

(913)768-1319
(800) 828-6474
Web site: http://www.nacanet.org
Bimonthly publication of the NACA

National Wildlife
National Wildlife Federation
11100 Wildlife Center Drive
Reston, VA 20190-5362
(703) 438-6000
(800) 822-9919
Web site: http://www.nationalwildlife.org/natlwild/index.html

Wildlife Control Technology Magazine
(815) 286-3039

ASSOCIATIONS

National Animal Control Association (NACA)
P.O. Box 480851
Kansas City, MO 64148
(913)768-1319
(800) 828-6474
Web site: http://www.nacanet.org

National Animal Damage Control Association (NADCA)
P.O. Box 2180
Ardmore, OK 73403

National Wildlife Control Operators Association
A&T Wildlife Management Services
1832 North Bazil Avenue
Indianapolis, IN 46219
(317) 895-9069
Web site: http://www.nwcoa.com

USDA National Wildlife Research Center
4101 LaPorte Avenue
Fort Collins, CO 80521
(970) 266-6000
Web site: http://www.aphis.usda.gov/ws/nwrc

11

WRANGLER

Wranglers work with horses. They groom, ride, train, lead pack trips or short trail rides, even round up cattle and other livestock. It's not a job for everyone—you have to be good riding on a horse, for one thing. You also have to love hard work and being outdoors all the time. But if you think you've got what it takes, wrangling is a very cool career. Just think, outside of the movies, how many real-life wranglers do you know?

Description

Wranglers do all kinds of work with horses. They work as grooms, caring for horses in their stables, saddling and unsaddling, and washing and brushing. They feed the horses hay and grain and give them plenty of water to drink. They shovel manure from stalls, organize tackle like saddles and bridles, and maintain horses' living areas, whether it's the ring, the barn, or an open field surrounded by fencing.

Wranglers also train horses. They work with them to make sure they'll do their job. This involves riding them every day, teaching them to be comfortable with a saddle, and teaching them how to walk, trot, canter, gallop, and stop on command. This is difficult work, especially with young horses!

Some wranglers lead organized pack trips into wilderness areas. They usually work for companies that specialize in outdoor adventures. They pack saddlebags with food and camping gear and load them onto the horses. Then they introduce

A wrangler takes novice riders on a trail ride.

their clients to their horses, teach them how to steer and stop if they're new to riding, and head for the hills. While on the trip the wranglers are not only responsible for the horses, but they also have to take care of their clients' needs—sometimes including setting up camp and cooking.

Other wranglers lead shorter trips called trail rides. These trips are many people's first experience with horses. As a wrangler leading a trail ride, you have to be able to control several horses at once. If your clients have never ridden before, they're relying on you to make sure they're safe for the entire trip. Often trails are rough and rocky, difficult to manage on foot, never mind on a horse! As a wrangler, you have to lead the way.

Wranglers employed at working ranches use their horses to round up cattle, sheep, and other horses. They usually work with a partner and sometimes specially trained dogs to team up and herd the animals in one direction. This is necessary when it's time to move a herd of cattle from one field to the next, or bring sheep in for shearing. Wranglers on ranches also ride around the perimeter of the property to check fencing. When they find a broken rail or post they stop and repair it.

Whatever job a wrangler does, you can be sure it involves a whole lot of hard work. Horses get sweaty, dirty, and dusty, and it's the wrangler's job to clean them. The weather gets

A wrangler herds wild horses.

finicky—the wrangler works rain or shine. Customers ask questions and need instruction—it's the wrangler's job to help them. Sound like fun? Grab a hat and a pair of cowboy boots and give it a try!

Education and Training

If you want to become a wrangler the first step is to learn how to ride. Take lessons at a local stable if you can. Spend as much time riding horses as possible. Ask lots of questions of your instructor and find out just what it takes to get good. And don't get frustrated in the process. Learning to ride a horse is not easy. It take time and practice.

In many cases, to be a good wrangler you also have to learn to work with people. This comes with practice. Learn to listen closely to people when they talk to you. See if you can explain things to people that they may not be familiar with. If you can get good at this, you'll have a much easier time teaching a new rider how to control a horse.

Many wranglers start out working on a farm or in a barn. They get experience cleaning stalls and being around horses before they move up to training horses and teaching others to ride them. It's usually even further on in their career that they actually use horses in working situations, on ranches, for instance.

Outlook

If you have excellent horse and people skills, and you know where to look, getting a job as a wrangler is not too hard. Many ranches, outfitters, and barns hire wranglers all the time. Just don't expect to make a lot of money. You're a cowboy or cowgirl, after all, not in the Kentucky Derby. Sarah of Caspar, Wyoming, says:

> I was a wrangler at a Wyoming lodge that took guests on trail rides through the woods. It was a

great job. I was able to spend time in a beautiful forest with amazing views, ride horses, and meet new and interesting people from all over the world. My qualifications for getting hired were good riding skills and good people skills.

There were times when I took people on trail rides who had never ridden before. That's a challenge! You have to be able to communicate well and teach new skills quickly and efficiently, whether to the elderly, small children, rowdy know-it-alls, or shy people. It is also important to have a good sense of safety and judgement since many times you are out in the woods and help is far away. You can't just leave your group of riders and run off to get help when a problem comes up.

Getting started as a wrangler is easy if you

A cowgirl leads a horse into a stable.

know how to ride, if you like people, and if you like to have fun. There are plenty of ranches around the country that like to hire young and energetic people. The money is good, too, as long as your clients have fun. Tips are the wrangler's bread and butter!

FOR MORE INFORMATION

BOOKS

Hill, Cherry. *Horse Care and Grooming: A Step-by-Step Photographic Guide to Mastering over 100 Horsekeeping Skills.* Pownal, VT: Storey Communications, 1997.

Vogel, Colin. *The Complete Horse Care Manual.* New York: DK Publishing, 1995.

WEB SITES

EquiSearch
http://www.equisearch.com

HorseClick
http://www.horseclick.com

HorseCountry.com
http://www.horsecountry.com

The Michigan Horse Page
http://www.horsepage.com

MAGAZINES

The American Quarter Horse Journal
P.O. Box 200
Amarillo, TX 79168
(806) 376-4811

Equus
P.O. Box 420235
Palm Coast, FL 32142-0235
(800) 829-5910

Horses All
North Hill Publications
4000-19 Street, NE
Calgary, AL T2P 2G4
Canada

Horse & Rider
1597 Cole Boulevard, Suite 350
Golden, CO 80401
(303) 445-4700
Web site: http://www.horseandrider.com.

The Horse: Your Guide to Equine Health Care
The Blood Horse, Inc.
P.O. Box 4680
Lexington, KY, 40544-4680
(606) 276-6771
Web site: http://www.thehorse.com.

Practical Horseman
P.O. Box 420235
Palm Coast, FL 32142-0235

(877) 717-8929

The Western Horseman
Western Horseman, Inc.
P.O. Box 7980
Colorado Springs, CO 80933-7980
(719) 633-5524

12

ANIMAL PHOTOGRAPHER

Would you consider yourself creative? Do you have a talent for seeing things just a little bit differently than everyone else? If so, and you also happen to love animals, photography could be just the thing for you. Animal photographers specialize in taking pictures of animals. Some might only shoot domestic animals like cats and dogs, while others stick to wildlife or show animals or sea life. Still others are generalists—they shoot photos of any animals they can, just as long as they have somewhere to sell those photos when they're done.

Description

Imagine that you are far out at sea on the SS *Reef Otter,* a large research vessel designed for scientists who study marine life. You're a self-employed, freelance underwater photographer on assignment for *National Geographic* magazine. Your mission is to follow the scientists everywhere they go and take photos of them and the animals they encounter. High on the waves, you strap on your scuba gear. You watch as two researchers disappear beneath the surface. You take a few deep breaths, check your gear— your special camera, your lighting equipment, your gauges and valves—and plunge over the side and into the water. You're off to work.

Animal photographers work in all types of environments. They snap shots of lizards and snakes in the Southwest desert, majestic bald eagles along the shores of

A wildlife photographer shoots seals in the Galapagos Islands.

Alaska, breaching whales in the middle of the ocean, and lions and hippopotamus in the jungles of Africa. They also do portraits of dogs and cats for their owners and document the happenings at circuses and zoos. As a matter of fact, there are photographers for just about every kind of animal—from ants to zebras!

Photography is an art. And good photographers spend huge amounts of time perfecting their work. They'll shoot dozens of rolls of film every day and spend countless hours developing that film and working in the lab to make prints. If they're lucky, they'll have assistants who can do some of the work for them.

Education and Training

Although a good creative eye is important in photography, so are many other skills. You need to know how to operate a professional-quality camera. You have to know all about apertures, film speeds, lenses, and lighting. You need to know about the entire photographic process, from positioning of the camera before the shot is taken to framing the finished product or selling the image to a buyer.

You can learn many of these skills in a photography class. Look for classes offered at your high school or local community college. Another way to learn is by just shooting pictures. Get a camera and go outside and shoot.

A photographer takes a picture of two blue-footed boobies perching on a rock in the Galapagos Islands.

See what it takes to get close enough to a bird to take its picture. Practice your angles. Take pictures of your pets. You won't believe how hard it is to make a pet look interesting in a photograph.

In addition to basic photography skills, you should also be good with animals. If you know what kinds of animals you want to specialize in, you should learn everything you can about them. If you want to shoot dogs, for instance, read up on different breeds, what they are meant to look like, and their habits. When it comes time to shoot, you'll

An underwater photographer with a manatee

know how to get just the right image to impress the owner or whomever you want to sell the picture to.

Most professional photographers get started by working at a newspaper or as an assistant to someone who is already established in the field. When you're ready to work, see if you can land a position at your local paper. Even if you don't shoot animals, the experience will set you up nicely for the future.

Outlook

It's tough to make a living as a professional photographer. It's one of those jobs that almost everyone would love to have, but few have the courage to try and even fewer have the good fortune to succeed. Of course, if you're hired by a wildlife magazine as a staff photographer, you'll never have a shortage of work. But if you work for yourself—freelance—you may find that work comes and goes. Sometimes you'll have more than you can handle, while at other times you'll have none at all. The key to success is sticking with it. With a good deal of persistence and a touch of talent, you'll find life as an animal photographer to be, well, picture perfect.

FOR MORE INFORMATION

BOOKS

Angel, Heather. *Natural Visions: Creative Tips for Wildlife Photography.* New York: Amphoto Books, 2000.

Benvie, Niall. *The Art of Nature Photography: Perfect Your Pictures In-Camera and In-Computer.* New York: Amphoto Books, 2000.

Edge, Martin. *The Underwater Photographer.* Boston, MA: Focal Press, 1999.

Havelin, Michael. *Practical Manual of Captive Animal Photography: The Step-by-Step Guide to Photographing Wildlife in Zoos, Aquaria, and Other Controlled Habitats.* Amherst, NY: Amherst Media, 2000.

Holmes, Judy. *Professional Secrets of Nature Photography: Essential Skills for Photographing Outdoors.* Buffalo, NY: Amherst Media, 2000.

WEB SITES

Nature Wildlife Photography Gallery
http://www.nature-wildlife.com

photo.net
http://www.photo.net

Photo Resource Magazine
http://www.photoResource.com

MAGAZINES

National Geographic
National Geographic Society
1145 17th Street NW
Washington, DC 20036-4688
Web site: http:///www.nationalgeographic.com

Outdoor Photographer
Werner Publishing Corporation
12121 Wilshire Boulevard, 12th floor
Los Angeles, CA 90025-1176
(310) 820-1500
Web site: http://www.outdoorphotographer.com.

Popular Photography
1633 Broadway
New York, NY 10019
(212) 767-6000
Web site: http://www.popphoto.com

Shutterbug Magazine
Patch Communications
5211 South Washington Avenue
Titusville, FL 32780
(321) 269-3312
Web site: http://www.shutterbug.net

Today's Photographer International
P.O. Box 777
Lewisville, NC 27023-0777
(336) 945-9867
Web site: http://www.aipress.com

ASSOCIATIONS

American Society of Media Photographers
150 North 2nd Street
Philadelphia, PA 19106
(215) 451-2767
Web site: http://www.asmp.org

Canadian Association of Photographers and Illustrators in Communications
100 Broadview Avenue, #322
Toronto, ON M4M 3H3
Canada
(416) 462-3700
(888) 252-2742
Web site: http://www.capic.org

North American Nature Photography Association
10200 West 44th Avenue, Suite 304
Wheat Ridge, CO 80033-2840
(303) 422-8527
Web site: http://www.nanpa.org

Professional Photographers of America, Inc.
229 Peachtree Street NE, Suite 2200
Atlanta, GA 30303
(404) 522-8600
Web site: http://www.ppa.com

GLOSSARY

apprenticeship A job training period working with someone who is more experienced.

career A profession or permanent job.

certification Formal recognition for having completed special training and having met certain qualifications.

diploma A certificate that recognizes completion of and graduation from school.

employee   One who works for an employer.

equine   Having to do with horses.

euthanasia   Killing or permitting the deaths of very sick or injured animals so that their suffering may end.

freelance   When a person works at a profession without a long-term commitment to any one employer.

humane   Friendly and compassionate consideration for animal welfare.

industry   A group of businesses in the same field.

labor   Work.

nonprofit organization   An organization that does not make a profit.

professional   A person with a career.

résumé   A summary of work experience, education, and skills.

salary   A predetermined amount of money to be paid to an employee for a year's work.

self-employed   Working for oneself, not for an employer.

technician   A specialist in the technical details of an occupation.

U.S. Bureau of Labor Statistics (BLS)   The government agency that keeps track of job trends in the United States.

volunteer   One who willingly works without pay.

wildlife refuge A place where wildlife can safely find shelter.

wildlife rehabilitation center A place where sick or injured wildlife are cared for and brought back to health.

work ethic How hard a person works at his or her job.

INDEX

About the Author

Chris Hayhurst is a freelance writer and photographer who specializes in the outdoors, sports, and environmental issues. In his spare time he enjoys hiking, rock climbing, telemark skiing, and anything that takes him into the backcountry. He lives in Santa Fe, New Mexico.

Photo Credits

Cover © Ron Garrison/AP Wide World/San Diego Zoo; pp. 9, 11 © Joe Raymond/AP Wide World; pp. 12, 13, 14 courtesy of Susquehanna Service Dogs; pp. 22, 24 © Earl Neikirk/AP Wide World; p. 25 © Syracuse Newspapers/The Image Works; p. 26 © Ian Stewart/AP Wide World; pp. 33, 34 © Erik Petersen/AP Wide World/*The Livingston Enterprise*; p. 37 © Kit Houghton Photography/Corbis; p. 39 © Anna Mia Davidson/AP Wide World; pp. 46, 51© Jay LaPrete/AP Wide World; p. 48 © Brad Armstrong/AP Wide World/*The Tribune*; p. 49 © Kathy McLaughlin/The Image Works; pp. 56, 59 © Jeff Kiessel/AP Wide World/*Ludington Daily News*; p. 60 © Mathew Grotto/AP Wide World/*Daily Southtown*; pp. 62, 98 © Paul A. Souders/Corbis; pp. 67, 68 © K.D. Lawson/AP Wide World; p. 70 © Orlin Wagner/AP Wide World; pp. 75, 80 © Kwame Zikomo/SuperStock; p. 78 © Omni Photo Communications, Inc./Index Stock Imagery; pp. 85, 89 © Grantpix/Index Stock Imagery; p. 87 © Jeff Robbins/AP Wide World; pp. 94, 96 © Sean Cayton/The Image Works; p. 104 © Earl Nottingham/AP Wide World/Texas Parks and Wildlife; p. 106 © Nancy Palmieri/AP Wide World; p. 108 © Robert Meccea/AP Wide World; pp. 102, 112 © Will Shilling/AP Wide World; pp. 117, 119 © Charles P. Saus/AP Wide World; p. 121 © Jeff Vanuga/Corbis; p. 123 © Richard Smith/Corbis; pp. 127, 131 © Robert Holmes/Corbis; p. 129 © Joel W. Rogers/Corbis; p. 132 © Brandon D. Cole/Corbis.

www.ingramcontent.com/pod-product-compliance
Lightning Source LLC
Chambersburg PA
CBHW050907210326
41597CB00002B/58